Engineering Geology

Edited by Essa Lwisa and Hasan Arman

Published in London, United Kingdom

IntechOpen

Supporting open minds since 2005

Engineering Geology
http://dx.doi.org/10.5772/intechopen.78106
Edited by Essa Lwisa and Hasan Arman

Contributors
Essa Georges Lwisa, Leonid Bogomolov, Lidia Sim, Pavel Kamenev, Chrysothemis Paraskevopoulou, Yacine Berrah, Nouar Charef, Serhane Brahmi, Abderrahmane Boumezbeur, Sylvestre Martial Ntomba, Bisso Dieudonné, Magnekou Takamte Rufine Christelle, Joseph Mvondo Ondoa, Attah Fakeye, Olusegun Ige, Olufemi Ogunsanwo

First published in London, United Kingdom, 2021 by IntechOpen
IntechOpen is the global imprint of INTECHOPEN LIMITED, registered in England and Wales, registration number: 11086078, 5 Princes Gate Court, London, SW7 2QJ, United Kingdom
Printed in Croatia

British Library Cataloguing-in-Publication Data
A catalogue record for this book is available from the British Library

Additional hard and PDF copies can be obtained from orders@intechopen.com

Engineering Geology
Edited by Essa Lwisa and Hasan Arman
p. cm.
Print ISBN 978-1-83881-864-7
Online ISBN 978-1-83881-894-4
eBook (PDF) ISBN 978-1-83881-898-2

For EU product safety concerns:
IN TECH d.o.o., Prolaz Marije Krucifikse Kozulić 3, 51000 Rijeka, Croatia, info@intechopen.com or visit our website at intechopen.com.

Meet the editors

Dr. Essa Georges Lwisa is an expert in core analysis, rock properties, formation evaluation, and enhanced oil recovery. He has worked at the Chemical and Petroleum Engineering Department, United Arab Emirates University since 2009 as a core analysis lab engineer. He is a reviewer for the *Journal of Petroleum Science and Engineering* and an editorial board member for several other journals. He has edited one book and written several book chapters.

Dr. Hasan Arman has been a professor in the Geology Department at United Arab Emirates University, College of Science, since 2008. He has been head of the department since August 2018. He received his BSc and MSc from Hacettepe and Istanbul Universities, Turkey, in 1984 and 1986, respectively. He obtained a Ph.D. from the University of Arizona, USA, in 1992. He worked as a postdoc at the University of Nevada, Reno, USA, from 1992 to 1993. He worked as a researcher at Tokai University, Japan, from 1995 to 1997 on a Japanese Monbusho Scholarship. He was also a faculty member at the Civil Engineering Department, Sakarya University, Turkey, between 1993 and 2008. Dr. Arman has taught several different courses at undergraduate and graduate levels. His research interests include rock mechanics, engineering geology, environmental degradation, sustainability, water resources, global warming, climate change, and renewable and sustainable energy sources. Dr. Arman has a number of publications in different peer-reviewed national and international scientific journals.

Contents

Preface

Engineering geology is the application of geology to engineering study for assuring that the geological factors regarding the location, design, construction, operation, and maintenance of engineering works are recognized and accounted for. Engineering geologists provide geological and geotechnical recommendations, analysis, and design associated with human development and various types of structures. The realm of the engineering geologist is essentially earth-structure interactions or investigation of how the earth or earth processes impact humanmade structures and human activities.

Engineering geology studies may be performed during planning, environmental impact analysis, civil or structural engineering design, value engineering and construction phases of public and private works projects, and during post-construction and forensic phases of projects. Works completed by engineering geologists include geological hazard assessments, geotechnical, material properties, landslide and slope stability, erosion, flooding, dewatering, and seismic investigations. Engineering geology studies are performed by a geologist or engineering geologist that is educated, trained, and has obtained experience related to the recognition and interpretation of natural processes, the understanding of how these processes impact humanmade structures (and vice versa), and knowledge of methods by which to mitigate hazards resulting from adverse natural or human-made conditions. The principal objective of the engineering geologist is the protection of life and property against damage caused by various geological conditions.

The practice of engineering geology is also very closely related to the practice of geological engineering and geotechnical engineering. If there is a difference in the content of the disciplines, it mainly lies in the training or experience of the practitioner.

In addition to a comprehensive introduction to engineering geology topics, this book focuses on subjects closely related to geomechanics, geotectonics, swelling soils, and dams.

Our thanks and gratitude go to Ms. Romina Rovan, Author Service Manager at IntechOpen, for her help and support during the work on this book.

Essa Lwisa
Department of Chemical and Petroleum Engineering,
United Arab Emirates University,
Al Ain, UAE

Hasan Arman
Department of Geology,
United Arab Emirates University,
Al Ain, UAE

Section 1

Introduction

Chapter 1

Introductory Chapter: Engineering Geology

Essa Georges Lwisa

1. Introduction

Engineering geology is the application of geology to the engineering study for the purpose of ensuring that the geological aspects related to the site, scheme, construction, process and maintenance of engineering works are recognized and taken in consideration [1].

Engineering geologists provide geological and geotechnical endorsements, analyzes, and designs related to human development and different kinds of structures. The field of engineering geology is primarily in the field of Earth-structure interactions, or the investigation of how Earth or Earth processes affect human structures and human activities [2].

Topics of Engineering Geology are:

- Rocks and geological structures: rock types, structures, plate tectonic. Engineering geological maps.
- Geo hazards: boundary hazards, ground subsidence, land slide, slope failure.
- Geological masses: mass fabric, ground mass description, weathering, rocks discontinuities.
- Field tests and measurements: tests in boreholes and excavations, engineering geophysics, seismic methods, electrical and magnetic methods.
- Ground improvement: shallow and deep impaction, grout treatment, bentonite suspension, ground anchor
- Water reservoirs and dams: dam design parameters, geological influences upon the selection of reservoir sites, dam foundations, dam seismicity.

2. Rocks and geological structures

Geologic structures are usually the result of the powerful tectonic forces that occur within the earth. These forces fold and break rocks, form deep faults, and build mountains. Repeated applications of force can create a very complex geologic picture that is difficult to interpret.

2.1 Rock types

Types of rocks are igneous, sedimentary and metamorphic. Igneous rocks are formed when molten rocks cools and solidifies. Sedimentary rocks arise when

IntechOpen

particles settle down out of water or air, or by precipitation of minerals from the water. Pile up in layers. Metamorphic rocks result when existing rocks are altered by heat, pressure, or reactive fluids, such as hot water rich in minerals [3].

2.2 Rock structures

Most rocks are not uniform all the time. On a scale best measured in millimeters or centimeters, they are made up of individual mineral grains that differ in size, shape and composition. The geometric properties of these small rock features and the relationships between them form rock texture. Rocks also commonly differ on larger scales, and are best measured in centimeters to meters to kilometers. The disparate and small-sized individual features of the rocks are called "structures". Our mission is to find out if there are rock structures that can provide clues to the formative environment of rocks: whether they are igneous, sedimentary or metamorphic.

There are many rock structures. Geologists usually divide them into "primary" and "secondary" structures [4].

- Elementary structures that were formed before or at the same time that matter is in the process of converting to rocks.

- Secondary Structures, imposed on rocks after they have already formed.

2.3 Plate tectonic

Plate tectonics is a scientific theory that describes the large-scale movement of seven large plates and the movements of many smaller plates from the Earth's lithosphere, since tectonic processes on Earth began between 3.3 and 3.5 billion years ago. The model is based on the concept of continental drift, an idea that developed during the early decades of the twentieth century. The theory of plate tectonics was accepted by the geological scientific community after sea floor propagation was validated in the late 1950s and early 1960s [5].

The lithosphere, the planet's rigid outermost layer, is divided into tectonic plates. The Earth's lithosphere consists of seven or eight main plates and many minor plates. When the plates encounter, their relative movement controls the type of boundary: convergent, divergent, or transformational. Earthquakes, volcanic activity, mountain building, and oceanic trench formation occur along these plate boundaries. The relative movement of the plates usually ranges from 0 to 100 mm per year [5].

2.4 Engineering geological maps

The Engineer Geological Mapping is a guide to the principles, concepts, methods, and practices involved in geological mapping, as well as applications of geology to engineering [6].

3. Geo hazards

Geographical hazards are geological and environmental conditions and involve long-term or short-term geological processes. Geographical risks can be relatively small features, but they can also reach huge proportions (for example, a landslide or submarine) and affect the local and regional social economy to a large extent (for example, a tsunami) [7].

3.1 Boundary hazards

There are three different types of converging plate boundaries recognized: continental, oceanic -oceanic, and continental - continental.

An oceanic ocean border is when two oceanic plates meet. Usually one plate submerges under the other and in the process a deep trench forms in the ocean and can also lead to the formation of undersea volcanoes [7].

A continental oceanic boundary is when an oceanic plate meets a continental plate and the denser oceanic plate descends below the continental plate [7].

The continental boundary is when two continental plates meet and neither of them subsides below the other because the continental rocks are relatively light and resist downward movement [7].

3.2 Ground subsidence

Subsidence is the sinking or settling of the Earth's surface. It can happen through a number of processes. Land subsidence may result from settling of local low-density soils, or the cavity of natural or man-made voids underground. Subsidence may occur gradually over many years as sagging or depressions form on the earth's surface. In rare cases, a sudden landing such as a dangerous ground hole may swallow any part of the structure in that location, or leave a dangerous, steep hole [8–10].

3.3 Land slide

A landslide may be defined as the movement of a mass of rock, debris, or land down a slope. Landslides are a kind of "mass wasting", which refers to any downward movement of soil and rocks under the straight influence of gravity. The word "landslide" includes five slope movement patterns: falls, slides, slides, spreads and flow [8, 11].

Landslides have several causes. Slope movement happens when forces acting on the lower slope surpass the force of the earth materials that make up the slope. Causes contain aspects that increase the effects of slope forces and aspects that contribute to a decrease or increase in strength. Landslides can start on slopes that are already on the edge of movement due to precipitation, snowmelt, changes in water level, table erosion, changes in groundwater, earthquakes, volcanic activity, disturbance caused by human activities, or any combination of these factors. Earthquake vibration and other aspects can also activate underwater landslides. These landslides are called undersea landslides. Sub-sea landslides sometimes cause tsunamis that damage seaside areas [9, 11].

3.4 Slope failure

Slope failure is the phenomenon of suddenly collapsing slope due to poor self-holding capacity of the earth under the influence of rain or earthquake. Due to the sudden collapse of the slope, many people fail to escape from it if it occurs near a residential area, resulting in a high death rate [12].

4. Geological masses

The size of the land that will be affected or will affect the engineering work. All rocks and many soil masses have discontinuities and their presence in rocks or soil mass is of prime importance for all engineering work in rocks or soil.

Mass movement, also called mass wasting, is the movements of soil and rock debris down slopes in response to gravitational pull, or the rapid or gradual sinking of the Earth's surface in a mostly vertical direction. Previously, the term mass wasting referred to a variety of processes by which large masses of cortical material are transported by gravity from one place to another. More recently, the term mass movement has been replaced to include processes of mass wasting and inundation of confined areas of the Earth's surface. The group movements on the ramps and the submersible group movements are often assisted by the water and the importance of both types is the role that each plays in changing the earthly shapes [13, 14].

4.1 Mass fabric

In geology, the texture of rocks defines the spatial and geometric formation of all the elements that make up it. In sedimentary rocks, the tissue developed depends on the deposition environment and can offer evidence on current developments at the time of precipitation. In structural geology, fabrics could deliver evidence on both the direction and size of strains that have controlled a particular piece of deformed rock.

Fabric types: [15, 16].

- Primary fabric
- Shape fabric
- Crystallographic preferred orientation
- S-fabric
- L-fabric
- Penetrative fabric
- Magnetic fabric.

4.2 Ground mass description

General term from the fine-grained, not discernible part of a rock. In igneous rocks, this is the part of the rock that is not phenocrysts, and can help in determining the composition of extrusive rocks. In sedimentary rocks, it typically refers to the fine-grained components, namely mud. In metamorphic rocks, it is usually referring to material between porphyroblasts or a low-grade rock with only microscopic mineralization [17].

A matrix or ground mass of rock is the mass of fine-grained substantial into which grains, crystals, or large holes are incorporated.

The matrix of igneous rocks contains of fine-grained, usually microscopic, crystals in which bigger crystals are fused. This porphyry tissue is revealing that magma was cooled in multi stages.

A sedimentary rock matrix is a fine-grained sedimentary material, such as clay or silt, in which larger grains or lumps are incorporated. It is also used to describe the rock material in which the fossil is included [7, 17].

4.3 Weathering

Weathering is the collapse of rocks on the Earth's surface by rainwater, temperature extremes, and biological activity. It does not involve removing rock material. There are three types of weathering, physical, chemical and biological [18].

4.4 Rocks discontinuities

Discontinuity in geotechnical engineering (in the geotechnical literature it is often referred to as a joint) is a surface or surface that indicates a change in the physical or chemical properties of a soil or rock mass. The discontinuity can be, for example, bedding, schistosomiasis, foliation, joint, splitting, fracture, cleft, crack or failed plane. Mechanical and integral discontinuities are separated. Interruptions may occur multiple times with the same mechanical properties on a large scale in a discontinuity group, or they may be a single interruption. The discontinuity causes the mass of soil or rock to anisotropy [7, 19].

5. Field tests and measurements

Geophysical surveys are primary sources for both qualitative and quantitative data regarding ground conditions, and they form an essential part of many on-site investigations. There are several reasons for this, perhaps the most important of which is that it provides, for design purposes, parameters that represent a more realistic assessment of geotechnical ground conditions than is usually the case with laboratory tests. The samples used for laboratory tests, due to their small size, may not be sufficiently representative of the ground from which they are taken. In particular, it may not have widespread discontinuities, found in rocks or soil masses, which greatly affect the engineering properties of the materials in question. Moreover, the sampling inevitably involves some disturbances in stress conditions and water content of soil and rocks so that the parameters obtained in the laboratory are not fully representative of the conditions at the site [20].

5.1 Tests in boreholes and excavations

Geotechnical investigations are performed by geotechnical engineers or engineering geologists to obtain information about the physical properties of soil works and proposed soil foundations for the proposed structures and to fix the distress of earthworks and structures caused by subterranean conditions. This type of investigation is called site inspection. In addition, geotechnical investigations are also used to measure the thermal resistance of soil or backfill materials required for underground transmission lines, oil and gas pipelines, radioactive waste disposal and solar thermal storage facilities. The geotechnical investigation will include surface exploration and subsurface exploration of the site. Sometimes, geophysical methods are used to obtain data about sites. Subsurface exploration usually includes soil sampling and laboratory testing of recovered soil samples.

Some of the on-site tests are: standard penetration test, dynamic cone penetration test, cone penetration test [7, 21].

5.2 Engineering geophysics

Engineering geophysics consists of the spatial studies of the Earth's surface and subsurface. The geophysical signal is measured, processed and analyzed in order to discover anomalies in the subsurface and determine the composition and physical properties of rocks, layers, etc. This information is essential in engineering planning, calculations and infrastructure project design, energy and environment [8, 22].

Use of geophysical methods to obtain information for civil engineering. The goal is usually to describe not only the geometry of the Earth's interior but also its nature (for example, its elastic properties as determined by measurements of seismic velocities and density). Shallow, gravitational, magnetic, and electrical seismic reflection and refraction methods and sampling methods are commonly used to find bedrock depth and sediment strength for foundation purposes, to determine the rupture (qv) susceptibility of rocks, to measure the degree of rupture, to detect underground cavities, to detect pockets of gas near the surface, to determine The dangers of buried pipelines under the sea floor, buried pollutant barrels, and landfill safety. In water-covered areas, high-powered sphygmomanometers, sparks, gas pistols, and other seismic reflection methods employ high frequencies (up to 5 kHz) to obtain reflections from shallow façades so that bedrock and the nature of the filling material can be diagnosed. Such methods are also used to locate large pipelines on the sea floor or to bury them on the sea floor by the prominent deflections they generate. It is usually limited to a shallow breakout of over 1,000 feet [22].

5.3 Seismic methods

Seismic tomography is a technique for imaging the Earth's interior with seismic waves produced by earthquakes or explosions. P and S waves and surface waves can be used for tomography models with different resolutions based on seismic wavelength, wave source distance, and seismometer array coverage [23].

Reflective seismology (or reflection seismic) is a technique of geophysics exploration that uses principles of seismology to approximation the properties of the Earth's interior from reflected seismic waves [24].

Seismic refraction is a geophysical standard governed by Snell's law of refraction. The seismic refraction method uses the refraction of seismic waves by rocks or soil deposits to characterize the subterranean geological conditions and geological building [24].

Seismic refraction is browbeaten in engineering geology, geotechnical engineering, and exploration geophysics. Seismic refraction traversal (seismic lines) are performed using a mixture of seismometers or geophones and a power source [24].

The procedures are based on the fact that seismic waves have different velocities in different types of soil or rocks. Waves are refracted when they cross boundaries between different types of soil or rocks. These methods allow the fortitude of general soil types and the estimated depth of stratigraphic boundaries or bedrock [24].

5.4 Magnetic and electrical methods

Magnetic techniques, including aeromagnetic surveys to map magnetic anomalies.

An aeromagnetic survey is a common type of geophysical survey that is performed with a magnetometer on board a plane or pulled behind it. This principle is similar to the magnetic survey conducted with a handheld magnetometer, but it allows for much larger areas of the Earth's surface to be quickly covered for regional reconnaissance. Plane typically flies in a grid-like pattern with line spacing and elevation to determine data accuracy (and cost per unit area of scan) [25].

Electrical techniques, including electrical resistivity tomography and induced polarization [26].

Stimulated polarization (IP) is a geophysical imaging technique used to determine the electric chargeability of subsurface materials, such as ores [27, 28].

Konrad Schlumberger originally discovered the effect of polarization when measuring the resistance of rocks [27, 28].

Induced polarization is a widely used geophysical method in mineral exploration and mine operations [27, 28].

IP scanning can be performed in both time area and frequency area mode [28].

Electromagnetic methods, such as magnetotellurics, ground penetrating radar, transient/time-domain electromagnetics and SNMR [29, 30].

Magnetotellurics (MT) is an electromagnetic geophysical method for inferring subsurface electrical conductivity from measurements of natural and geoelectric magnetic field anisotropy at the Earth's surface. Search depth ranges from 300 meters underground by recording higher frequencies up to 10 km or deeper with sounding for prolonged periods [29, 30].

Ground penetrating radar (GPR) is a geophysical method that uses radar pulses to photograph the Earth's interior. It is a non-intrusive way to survey below the surface to check underground facilities such as concrete, asphalt, minerals, pipes, cables or masonry [31, 32].

Transient electromagnetism, is a geophysical exploration method in which electric and magnetic fields are induced by transient pulses of an electric current and the subsequent decay response is measured [33, 34].

6. Ground improvement

Ground improvement refers to a technology that improves the engineering properties of a mass of treated soil. Usually the properties that are modified are shear strength, stiffness, and permeability. Floor improvement has evolved into a sophisticated tool to support foundations for a variety of structures [35].

6.1 Shallow and deep impaction

Soil compaction is the process in which pressure is applied to the soil causing concentration as air is exiled from the pores between the soil grains. The compaction is usually the result of heavy machinery compressing the soil [35, 36].

The available techniques can be classified as: Static, Impact, Vibrating, Gyrating, Rolling, and Kneading [35, 36].

6.2 Grout treatment

A ground remediation operation performed to accomplish one of two things, either to reduce water flow or to improve the properties of the ground by drilling wells in the foundation and injecting material under pressure into the subsurface foundation [35, 37].

Each hole in the filler project is an extension of previous exploration pits and the data collected is used to increase the understanding of subsurface conditions. Injectable materials used for filler range from cementitious plaster materials (particles) to a variety of chemical slurries [37].

Filling foundations in dams and canals to reduce water flow are among the oldest applications of fillers, dating back to the early nineteenth century with plaster curtains in use since the 1890s [37].

6.3 Bentonite suspension

The different kinds of bentonite are called after their main element, like potassium, sodium, calcium, and aluminum. Bentonite is usually made by weathering volcanic ash, regularly in the presence of water. However, the term bentonite, as well as a similar clay called Tonstein, has been used to describe clay layers of uncertain origin. For industrial purposes, there are two main classes of bentonite: sodium and calcium bentonite. In stratigraphy and tephrochronology, fully demixed ash layers (weathered igneous glass) are commonly referred to as K-bentonites when the predominant clays are lit. In addition to montmorillonite and the attachment of other common clays that are sometimes prevalent are kaolin. Clays dominated by kaolinite are commonly referred to as tonsteins and are commonly associated with charcoal [35, 38, 39].

6.4 Ground anchor

Ground anchors are a device designed to support structures, most commonly used in geotechnical and construction applications.. Ground anchors are used in temporary and permanent applications [35].

Ground anchors are commonly used in civil engineering and construction projects, and have a variety of applications, including: [35].

- Retaining walls.
- Structural support of temporary buildings and structures,
- Tethering marine structures.
- Supporting guyed masts.
- Anchoring utility poles.
- Landscape, anchoring trees, often semi-mature transplants.
- General security, as in anchoring small aircraft.
- Sporting activities, such as slacklining or abseiling.

7. Water reservoirs and dams

The reservoir is an non-natural lake in which water is stored. Most reservoirs are made by building dams across rivers. A reservoir can also be made from a natural lake whose outlet has been blocked to control the water level. The dam controls the amount of water that streams from the reservoir [40].

7.1 Dam design parameters

Design criteria for earth dams are: [41].

- Sufficient capacity is provided for the drainage and the float basin so there is no risk of overflow of the dam.
- The leakage flow across the bridge is controlled so the amount lost does not interfere with the target of the dam and there is no erosion or erosion of the soil. In this regard, the leakage line should remain well within the downstream front of the dam and the part of the weir should be drained on the downstream side of the impermeable core.
- The uplift pressure caused by the leakage from below is not enough to cause the pipes.

- Bridge slopes are stable under all tank operating conditions, including rapid drawdown and during continuous leakage under a full tank.

- The stresses imposed by the bridge on the foundation are less than.

- The upstream face is properly protected ((stone throw, riprap, revetment) from abrasion caused by the movement of waves, and the lower face (anti-arms, grass) is protected from the impact of rain.

7.2 Geological influences upon the selection of reservoir sites

Topography: In the geological sense, topography is the composition of the Earth's surface, and includes the location, size and shape of physical features such as hills, hills, valleys, streams, and lakes. Topographic maps show these features. Examination of a topographic map combined with a survey of the land is often sufficient to determine the overall topographical suitability of the dam and the location of the proposed reservoir. This is the first and easiest step in determining the feasibility of a proposed project [25].

Hydrology and Hydrogeology: Hydrology is the science related to the Earth's water, its distribution and phenomena. For a dam and reservoir project to be successful, it must have an adequate and continuous supply of water suitable for the tank's intended uses. Hydrological information and investigations will be required to varying degrees, depending on the size of the project. Annual rainfall, the ratio of the catchment area to the reservoir area, and the size of the stream flow must be known in all seasons of the year. It is also necessary to study groundwater science to determine whether the groundwater will contribute to the reservoir or whether the reservoir will lose water to the groundwater system. Tank capacity, maximum and minimum tank yield must also be known so that water commitments do not exceed the amount of water available [25].

Geology: To properly judge the feasibility of the proposed dam and reservoir project, it is necessary to know the type, distribution and succession of rocks and other geological units in the project area, for the stability of the dam and water - the ability to maintain the reservoir is directly related to them. The aspects of geology that must be evaluated to determine the suitability of a project site include:

- The directions of the units, whether flat or tilted.

- The depth and extent of weathering.

- The presence and condition of breaks, such as open or closed joints, faults, or solution channels.

- The presence of layers of sand or silt and old soil areas.

The engineering properties of the geological units are directly related to the type of rocks or unconsolidated material involved, and thus to the geology [25].

7.3 Dam foundations

A site investigation shall be carried out prior to construction to verify the nature of the foundation. By knowing the actual foundation condition at the site, the earth dam can then be designed accordingly. An embankment foundation is said to be suitable if it is able to provide stable support for the bridge under all conditions of saturation and loading and that it provides adequate leakage resistance to avoid excessive water loss [42, 43].

The foundation of the dam may be broadly classified into three types which are rock foundations, coarse-grained material foundations and fine-grained material foundation [43].

These foundations may need to be treated to stabilize any weakness and also to reduce leakage. On the other hand, rock foundation must be inspected for erosive leakage and excessive uplift pressure. If such conditions exist, the foundation must be considered grouting [9, 43].

7.4 Dam seismicity

Induced earthquakes refer to the earthquakes and slight tremors that result from human activity that alter the stresses and stresses on the Earth's crust. Most of the induced earthquakes are of low magnitude [44].

Seismic hazard from induced seismic activity can be assessed using techniques similar to natural earthquakes, although one account for unstable earthquakes. Earthquakes vibrating from induced earthquakes appear to be similar to those observed in natural tectonic earthquakes, although differences in rupture depth need to be taken into account. This means that ground motion models derived from natural seismic recordings can be used, which are often more numerous in robust motion databases compared to induced earthquake data. Then, a risk assessment can be performed, taking into account earthquake risk and the vulnerability of vulnerable items (such as local residents and building stock). Finally, risk can, in theory at least, be mitigated, either through modifications of the risk or reduced exposure or vulnerability [44].

Acknowledgements

All thanks and gratitude are for my wife, Rasha, and my children, George Alexander and Tia, for their unconditioned love.

I am also grateful for Prof. Hasan Arman, Prof. Sulaiman Alzuhair, and Prof. Ali Almarzouqi, from UAE University for their support.

Author details

Essa Georges Lwisa
Chemical and Petroleum Engineering Department, United Arab Emirates University, Al Ain, UAE

*Address all correspondence to: essa.lwisa@outlook.com; essa.lwisa@uaeu.ac.ae

References

[1] B. a. Jacson, Glossary of Geology, AGI, 1981.

[2] Kiersh, "The Heritage of Engineerng Geology: The first 100 years," Geological Society of America, vol. 3, 1991.

[3] A. M. o. N. History, "Exhabitions," Amirecan Museum of Natural History, 6 1 2021. [Online]. Available: https://www.amnh.org/exhibitions/permanent/planet-earth/how-do-we-read-the-rocks/three-types.

[4] U. S. D. o. Agriculture, National Engineering Handbook, United States Department of Agriculture, 2012.

[5] R. F. Butler, Applications to paleogeography, IUPAC, 1992.

[6] W. Dearman, Engineering Geological Mapping, Elsevier Ltd., 1991.

[7] D. G. Price, Engineering Geology: Principles and Practice, Springer, 2008.

[8] A. a. farmer, Principles of Engineering Gelology, University of Durham, 1976.

[9] T. Waltham, Foundations of Engineering Geology, CRC Press, 2009.

[10] C. G. Survey, "Colorado Geological Survey," 12 12 2020. [Online].

[11] "U.S. Geological Survey," 12 12 2020. [Online]. Available: https://www.usgs.gov/faqs/what-a-landslide-and-what-causes-one?qt-news_science_products=0#qt-news_science_products.

[12] I. S. Network, "International SABO Network," [Online]. Available: http://www.sabo-int.org/dott/slope.html#:~:text=A%20slope%20failure%20is%20a,a%20higher%20rate%20of%20fatalities.

[13] d. F. M.H., Geological Masses, Springer, Berlin, Heidelberg, 2009.

[14] T. E. o. E. Britannica, "Mass movement," Encyclopaedia Britannica, 12 12 2020. [Online]. Available: https://www.britannica.com/science/mass-movement.

[15] M. W. &. W. P. Hobbs BE, An outline of structural geology, John Wiley & sons, 1976.

[16] T. R. a. M. EM, Structural Geology, WH Freeman and Co, 2007.

[17] M. D. A. P. I. C. M. Chris Johnson, an Introduction to Geology, Salt Lake Community College, 2017.

[18] T. G. S. o. London, "Weathering," The Geological Society of London, 12 12 2020. [Online]. Available: https://www.geolsoc.org.uk/ks3/gsl/education/resources/rockcycle/page3461.html#:~:text=Weathering%20is%20the%20breakdown%20of,%2C%20physical%2C%20chemical%20and%20biological.

[19] R. Ulusay and J. Hudson, The Complete ISRM Suggested Methods for Rock Characterization, Testing and Monitoring, Ankara: Turkish National Group, 1974.

[20] d. F. M.H., Field Tests and Measurements, Springer, Berlin, 2009.

[21] ASTM, "Standard Test Method for Standard Penetration Test (SPT) and Split-Barrel Sampling of Soils," ASTM International.

[22] T. S. o. E. Geophysicists, "Engineering geophysics," 12 12 2020. [Online]. Available: https://wiki.seg.org/wiki/Dictionary:Engineering_geophysics.

[23] A. Malehmir, M. Urosevic, G. Bellefleur, C. Juhlin and B. Milkereit,

Seismic methods in mineral exploration and mine planning — Introduction, The Society of Exploration Geophysicists, 2012.

[24] J. Milsom and A. Eriksen, Field Geophysics, Environmental and Engineering Geoscience, 2011.

[25] S. A. J. C. Burger RH, Introduction to Applied Geophysics, W. W. Norton, 2006.

[26] M. Loke and R. Barker, "Practical techniques for 3D resistivity surveys and data inversion," in the 57th EAEG Meeting, Glasgow, 1995.

[27] D. F. Bleil, "INDUCED POLARIZATION: A METHOD OF GEOPHYSICAL PROSPECTING," GEOPHYSICS, vol. 18, no. 3, pp. 636-661, 1953.

[28] J. W. S. U. D. L. C. J. H. C. K. S. C. R. H. C. L. S. E. P. G. P. G. H. J. D. K. T. B. M. D. H. D. P. L. F. D. M. M. H. L. R. D. B. S. C. M. Ken Zonge, "Resistivity, Induced Polarization, and Complex Resistivity," Investigations in Geophysics, pp. 265-300, 2005.

[29] A. Tikhonov, "On determining electrical characteristics of the deep layers of the Earth's crust," Doklady., vol. 73, no. 2, pp. 295-297, 1950.

[30] L. Cagniard, "Basic theory of the magneto-telluric method of geophysical prospecting," Geophysics, vol. 18, no. 3, p. 605-635. , 1953.

[31] D. DJ, "Ground Penetrating Radar," Institution of Engineering and Technology, no. 2, pp. 1-4, 2004.

[32] J.-F. Hofinghoff, "Resistive Loaded Antenna for Ground Penetrating Radar Inside a Bottom Hole Assembly," IEEE Transactions on Antennas and Propagation, vol. 61, no. 12, p. 6201-6205, 2013.

[33] M. Nabighian, "Quasi-static Transient Response of a Conducting Half-Space – an Approximate Representation," Geophysics, vol. 44, no. 10, pp. 1700-1705, 1979.

[34] J. D. McNeill, "Applications of Transient Electromagnetic Techniques," in Technical Note 7, Geonics Ltd, Mississauga, Ontario., 1980.

[35] A. B. Klaus Kirsch, Ground Improvement, New York: CRC Press, 2013.

[36] X. Jia, W. Hu, P. Polaczyk, H. Gong and B. Huang, "Comparative Evaluation of Compacting Process for Base Materials using Lab Compaction Methods," Journal of the Transportation Research Board, vol. 2673 , no. 4, p. 558-567, 2019.

[37] B. M. Peter T. Bobrowsky, "Grout/Grouting," Encyclopedia of Engineering Geology, pp. 1-15, Living Edition.

[38] I. E. Odom, "Smectite clay Minerals: Properties and Uses," Philosophical Transactions of the Royal Society A: Mathematical, Physical and Engineering Sciences., vol. 311, no. 1517, p. 391-409, 1984.

[39] K. Bekkour, "Rheological Characterization of Bentonite Suspensions and," Applied Rheology, vol. 11, no. 4, pp. 178-188, 2001.

[40] N. G. Editiors, "Reservoir," National Geographic, 12 12 2021. [Online]. Available: https://www.nationalgeographic.org/encyclopedia/reservoir/.

[41] G. Mishra, "CRITERIA FOR DESIGN OF EARTH DAMS," The Constructor - The Construction Encyclopedia, 12 12 2020. [Online]. Available: https://theconstructor.org/water-resources/criteria-for-design-of-earth-dams/2278/.

[42] H. R. H. R. G. P. S. Ö. Best E., "Development, Danube river," in Danube river: Development, Dordrecht, 1998.

[43] UKEssays, "Design and Construction Fundamentals Of Earth Dams Environmental Sciences Essay." UKEssays, 12 12 2018. [Online]. Available: https://www.ukessays.com/essays/environmental-sciences/design-and-construction-fundamentals-of-earth-dams-environmental-sciences-essay.php#citethis.

[44] R. S. J. C. J. H. E.M. Gosschalk, "An Engineering Guide to Seismic Risk to Dams in the United Kingdom, and its international relevance," Soil Dynamics and Earthquake Engineering, vol. 1, no. 3, pp. 163-179, 1994.

Section 2

Geomechanics

Chapter 2

Time-Dependent Behavior of Rock Materials

Chrysothemis Paraskevopoulou

Abstract

Understanding the geomechanical behavior of a geological model is still an on-going challenge for engineers and scientists. More challenges arise when considering the long-term behavior of rock materials, especially when exposed to environments that enable time-dependent processes to occur and govern overall behavior. The latter is essential in underground projects such as nuclear waste repositories. The lifespan can exceed one million years or other openings where the project's lifetime and sustainability are the critical design parameter. In such cases, progressive rock mass deformation that can lead to instabilities, time-dependent overloading of support and delayed failure are considered the product of time-dependent phenomena. Understanding and predicting the overall impact of such phenomena aims to achieve design optimization, avoiding dlivery delays and thus cost overruns. This chapter provides more insight into the time-dependent behavior of rocks. Simultaneously, the emphasis is given to investigating and analyzing creep deformation and time-dependent stress relaxation phenomenon at the laboratory scale, and in-depth analyses are presented. This work further develops the understanding of these phenomena, and practical yet scientific tools for estimating and predicting the long-term strength and the maximum stress relaxation of rock materials is presented. The work presented in this chapter advances the scientific understanding of time-dependent rock, and rock mass behavior increases the awareness of how such phenomena are captured numerically and lays out a framework for dealing with such deformations when predicting tunnel deformations.

Keywords: time-dependent behavior, long-term behavior, long-term strength, creep, stress-relaxation, strength-degradation

1. Introduction

Rock engineering and tunneling are considered to be three-dimensional problem. In practice, the short-term mechanical performance is of primary focus in design as design and characterization parameters and data are derived from short-term testing. Challenges and implications can be formed when performance over time and long-term behavior is taken into consideration. Current design methodologies used in underground structures and tunneling projects are commonly solely based on the static response of the surrounding ground neglecting the long-term time-dependent behavior that can affect the overall structure's performance and the construction process [1, 2]. The latter can cause difficulties when attempting to understand the governing mechanisms in rock materials where time-dependent phenomena such as creep and stress relaxation can occur [3, 4]. When these processes are excluded or neglected during the design process, incorrect results and unsound conclusions are derived. These can involve support requirements and excavation methods employed, impacting the construction, the maintenance cost of the tunnel, and in the worst case, may even cause safety issues [5–8].

Strength-degradation is considered highly important in underground applications such as low, intermediate and high-level nuclear waste. The time-dependent strength decrease deteriorates the overall lifetime of the underground opening [9]. This lifetime span can range from 100,000 to 1,000,000 years which significantly exceeds the typical 100-year lifetime of underground projects. It is evident, thus, the reason why there is a need to investigate from micro to macro-scale further the long-term behavior of rock materials that could be used as host-rocks for such applications.

This Chapter aims to provide more insight into rock materials' time-dependent behavior by addressing the mechanisms involved and highlighting the associated implications for both scientific and practical applications. In this work, both experimental laboratory testing and numerical analyses are employed to examine the time-dependent mechanisms and rocks' response under different boundary conditions while introducing a different perspective for analyzing and predicting the intact rock's time-dependent behavior of the rock mass behavior in underground environments. A time-dependent response such as creep, squeezing, swelling, stress relaxation, and strength degradation of the rock mass can occur during both the construction and the maintenance of underground openings depending on the in situ conditions that control the mechanical behavior shown in **Figure 1**.

It has been observed that an often misconception is the assumption that time-dependent phenomena only act individually. However, this assumption can yield unsound estimations and erroneous conclusions. These phenomena may share the same (or similar) mechanisms given the existing in situ conditions can take place either in series or even simultaneously. Therefore, the overall observed displacement on the tunnel wall can result from different phenomena acting together. The selection of an appropriate constitutive model to examine the mechanical behavior of rock material overtime is required. The ability of such models to capture and simulate time-dependent behavior is illustrated in **Figure 2**.

The time-effect can cause different behavioral patterns depending on the underground construction project's site-specific conditions; the selection of the

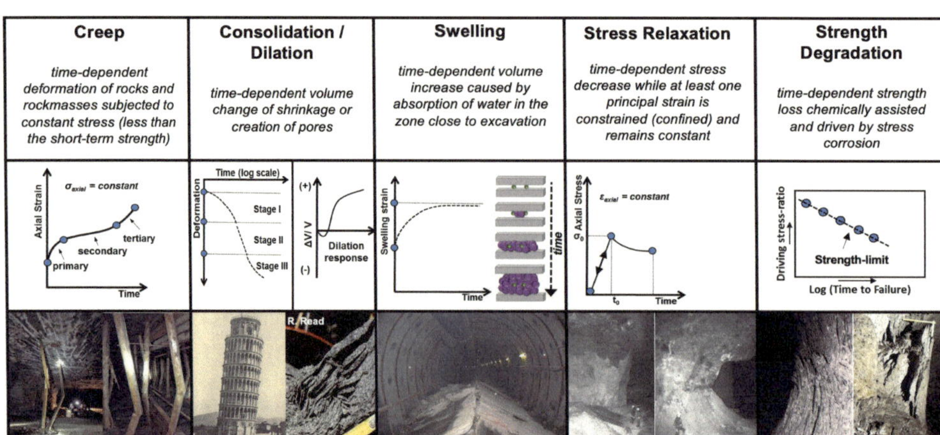

Figure 1.
Examples of time-dependent phenomena, the behavioral response with time and a description of the phenomena encountered in rock tunneling.

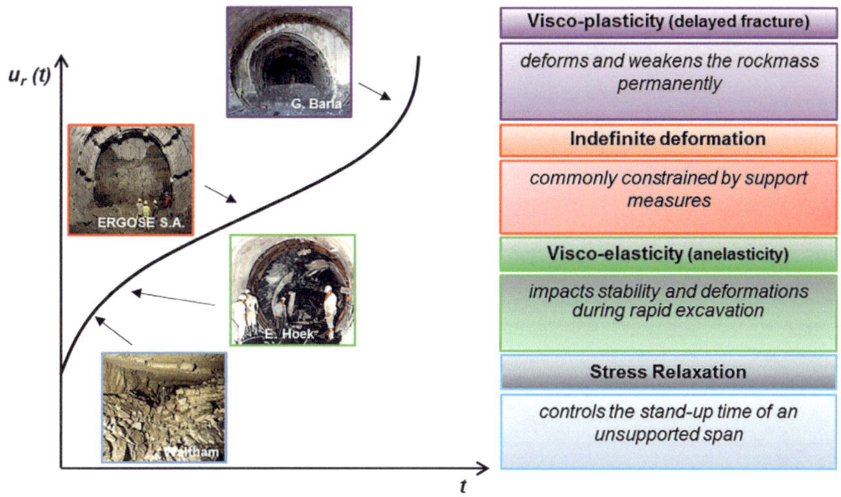

Figure 2.
Examples of reported failures and mechanisms associated with time-dependent behavior; where t refers to time and ur(t) to the radial displacements observed in the tunnel walls over time.

appropriate model to simulate the desired mechanical response is crucial. For instance, stress relaxation usually occurs near the newly exposed walls after the tunnel face excavation. The visco-elasticity (or anelasticity) can cause implications during rapid excavation (i.e., TBM). The indefinite deformation usually observed in more ductile materials can result from the deterioration of the support system. In this case, different support measures (i.e., yielding support systems) should be undertaken where the on-going deformation will be allowed to take place. Visco-plasticity or delayed fracturing can permanently damage the rock mass after initial construction, requiring redesigning the initial tunnel design. These examples show the importance of using the appropriate model to simulate the real conditions as closely as possible and estimate how the rock mass will behave over time.

2. Theoretical and practical background

Different rocks and rock masses respond in different ways over time. The main factor that controls their behavior is geology. The mineralogical content and the geological structure impact rocks' mechanical behavior; ultimately, the stress regime and the environmental conditions also influence the rock materials' behavior. **Figure 3** provides a roadmap on the material's anticipated mechanical behavior grouped into ductile or brittle behavior based on the conditions the material is initially formed. In general, as the temperature and confining pressure increase, the rock transitions from brittle to ductile (**Figure 3a**). Brittle materials tend to abruptly fail as the stress approaches their short-term strength, and as such, they absorb less energy. In contrast, ductile materials can sustain an applied stress state through more deformation (**Figure 3b** and **c**). When ductile materials (i.e. rock salt or potash) are subjected to constant differential stress below their nominal yield strength, they can behave as visco-elastic materials and further deform as time elapses (**Figure 3d** and **e**). In contrast, brittle materials (i.e., granite or limestone) under similar stress conditions may only exhibit micro-crack damage with progressive crack propagation that results in the eventual interaction of the previously isolated microcracks, which leads to sudden failure (**Figure 3d** and **e**).

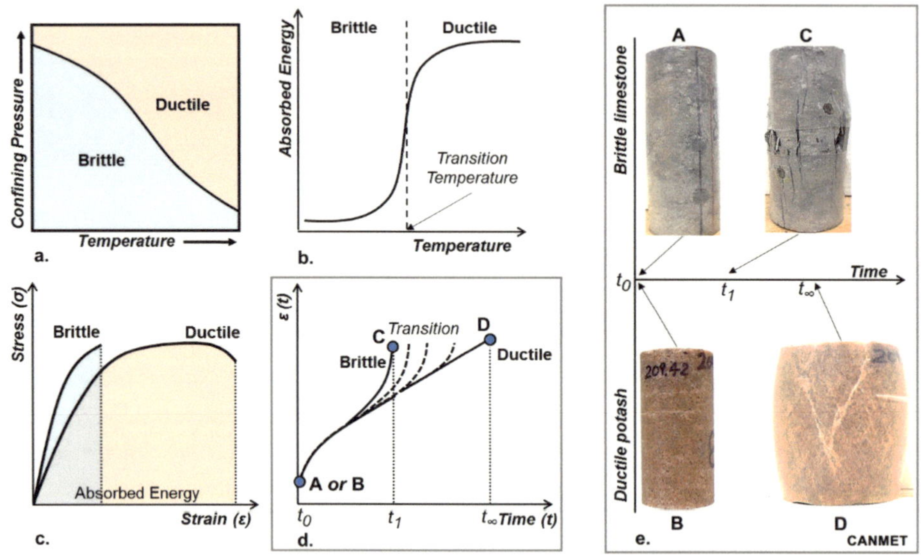

Figure 3.
Schematic illustration and comparison between brittle and ductile rock materials, (a) transition from brittle to ductile behavior according to confining pressure and temperature conditions; (b) absorbed energy and temperature; (c) general stress – Strain behavior of brittle and ductile materials; (d) stain- rate and time relationship of brittle and ductile materials subjected to constant stress exhibiting creep, and (e) examples of brittle limestone and ductile potash before and after static load (creep) tests.

2.1 Time-dependent phenomena

Time-dependency refers to the deformation of rock (or other materials) over time. Mechanisms deforming or weakening the rock mass over time are called time-dependent phenomena. Since the late 1930s, researchers started investigating the effect of time in rock behavior, trying to apply the theory of creep widely studied and reported on metals [10] to rock behavior. It was not until 1939 when Griggs [11] undertook laboratory experiments to examine the phenomenon of creep of rocks. He constructed two apparatus and performed tests on limestone, anhydrite, shale and chalk. He also examined recrystallization under creep conditions at high pressure. At the excavation scale, addressing the effect of time in tunneling and mining engineering has been studied since the 1950s. Researchers introduced the idea of 'stand-up time' in tunnel stability. The 'stand up time', a reflection of time-dependent weakening, was also included in the rock mass classification systems [12–14], emphasizing time and its effects by producing charts illustrating the time frame of stable unsupported spans. Since the 1960s many researchers [15–25] have investigated the influence of time on the long-term strength of rock by performing laboratory testing on rock samples, typically using static load (creep) tests by sustaining a constant stress condition. Creep phenomenon is most commonly applied to the study of soft, mono-mineralic rocks such as halite, potash, and limestone [26]. Following this practice, new constitutive and numerical time-dependent models were introduced based on the experimental results and data [27–31]. These models attempt to capture and reproduce the behavior of laboratory tests on the rocks, including time.

In practice, as previously mentioned, there is often a miscomprehension and misinterpretation of the different time-dependent phenomena and the mechanisms acting and resulting in weakening rock and the rock mass over time [9]. This section serves as an attempt to redefine and describe the various mechanisms that can appear to be time-dependent under the appropriate conditions using the composite

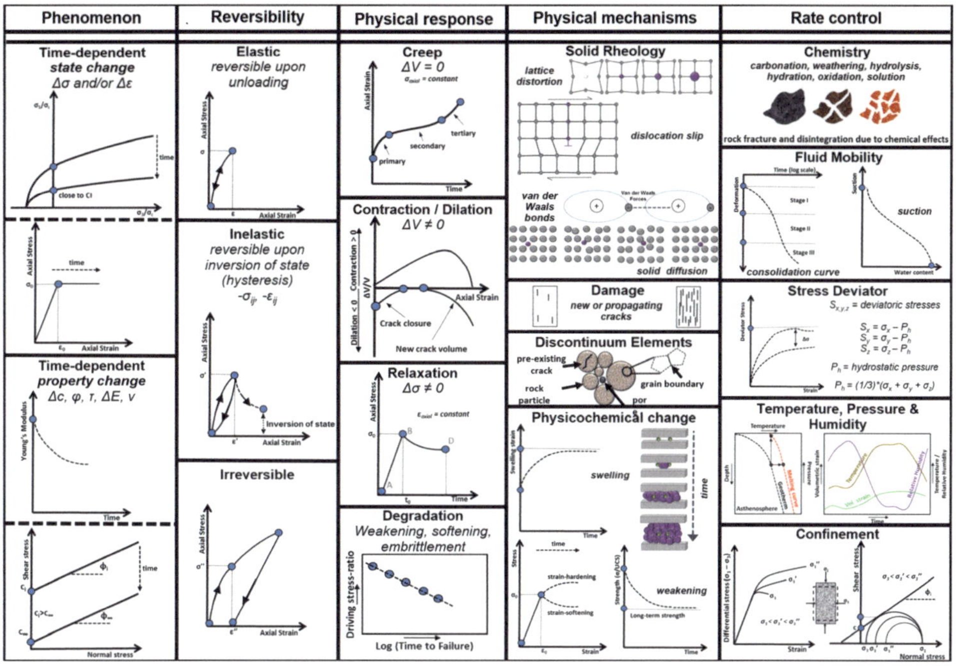

Figure 4.
Nomenclature, defining time-dependent phenomena and the conditions and mechanisms that affect and govern the rock behavior [9].

nomenclature shown in **Figure 4**. The phenomenon can be either due to a state-change (i.e. stress decrease) or a property-change (i.e. decrease in cohesion). These changes can be further categorized according to their reversibility or recoverability as elastic, inelastic, and irreversible and may increase to visco-elastic or visco-plastic strains. The physical response can be represented as creep (shear strain), contraction or dilation (volumetric strains) over time, as well as relaxation (reduction in shear stress under sustained strain) and degradation (strength loss) depending on loading and boundary conditions. The micro-mechanical mechanisms tend to vary according to the boundary conditions. For instance, the solid rheology (e.g. lattice distortion, dislocation slip, van der Vaal's bonds and/or solid diffusion) may be damaged by new cracks that initiate or pre-existing ones propagate while pores, grain boundaries, and pre-existing cracks creating discontinnuum elements. Besides, the physicochemical changes can be temporal, rheological, and chemical alterations in the micro-scale, leading to swelling, weakening, strain-softening, and hardening. The rate and the magnitude of the time-dependent performance of rock materials are controlled by other environmental, physical, and loading conditions (e.g. temperature, pressure, humidity, and confinement).

Time-dependent phenomena can be a combination of many factors that can result in various physical responses and act either simultaneously or individually. Differentiating and recognizing these phenomena can be a complicated process, and all components in **Figure 4** should be taken into account.

The overall physical response can be a combination/integration of the mechanisms that influence the long-term behavior of intact rock and rock masses and include:

- creep during which visco-elastic behavior governs where time-dependent, inelastic strains and 'indefinite' deformation occur and/or visco-plastic yield where time-dependent plastic strains occur that lead to permanent deformation.

- dilation or contraction where volume change takes place over time usually caused by the change of stress resulting in the propagation and interaction of cracks (dilation) or the closure of the existing ones (contraction).

- relaxation where the reduction of the stress with time under sustained strain is controlled by the internal creep processes aimed at relieving the stored elastic energy

- mechanical property degradation where strength and/or stiffness change due to damage processes that accompany or occur as a result of the above phenomenon.

2.2 Time-dependent laboratory tests

Time-dependent behavior of rock materials is usually investigated in the lab-scale by performing static load (creep) and stress relaxation tests which can be done in uniaxial and triaxial compressive conditions.

2.2.1 Relaxation tests

Relaxation is defined as stress (or load) decrease over time when the deformation (or strain) is kept constant. Commonly, the axis on which the stress is applied (i.e. axial stress used) determines the deformation's axes that are maintained constant (i.e. axial strain – constant). It has been observed that relaxation behavior is related not only to time-dependent phenomena like creep but also to time-dependent damage evolution of new or pre-existing cracks growth and evolution in the specimen that initiates during loading [9, 32, 33].

Figure 5 shows the stages during a stress relaxation test from A to C. The rock is initially loaded in the axial direction up to point A, which is considered the strain threshold at which the applied strain is held constant (points A to C). In this regard, these tests are often referred to as strain-controlled. Overtime, existing cracks and/or new cracks are formed and propagated at this strain threshold, contributing to the observed stress decrease (relaxation). When this stress relaxation reaches an asymptote (no further decrease is observed), the test is terminated, which implies that crack growth stabilization is achieved [19].

It should be stated that suggested standard test guidelines on relaxation tests on rock samples are not provided by ISRM. However, there are guidelines provided by ASTM [34] for relaxation testing performed on man-made materials and structures. In section 3.2 this standard has been adopted and adjusted for rock relaxation testing.

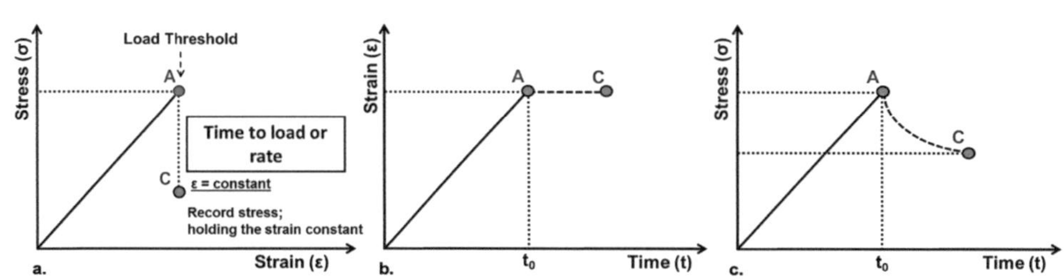

Figure 5.
Relaxation test: (a) stress–strain response, (b) strain- time response, (c) stress-time response.

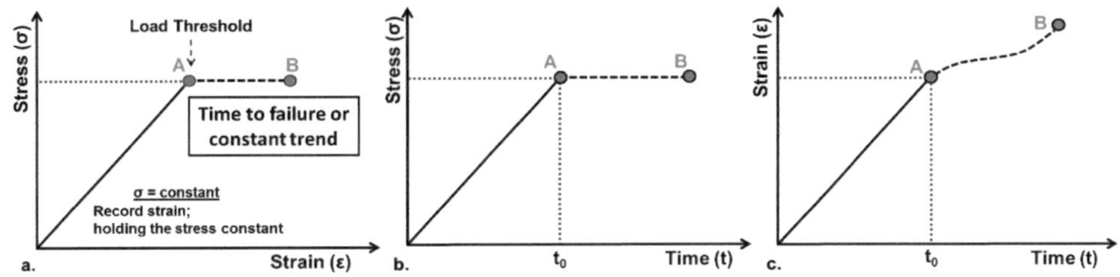

Figure 6.
Static load test: (a) stress–strain response, (b) stress-time response, (c) strain-time response.

2.2.2 Static load tests

To investigate creep, this time-dependent deformation of materials subjected to constant load or stress less than its short-term strength, static load tests are performed. In materials, here is a minimum load or stress, which enables them to undergo creep behavior, below which no creep is observed [9, 35, 36]. Elevated differential stress triggers the deformation of crystal lattices, leading to straining of the minerals, potentially microcracking, and eventually measurable strain of the rock element. **Figure 6** presents a typical stress–strain-time response of a (uniaxial) creep test. The rock sample is loaded until point A, the stress threshold where it is held constant. Over time the strain increases at different rates up to point B, where failure occurs. This test is usually referred to as load-controlled or stress-controlled tests.

Failure of the specimen usually denoted the completion of the test. However, many static load tests are terminated when a constant strain-rate is achieved, inferring the transition to the secondary stage of creep. For static load tests, ISRM [37] has suggested standard guidelines.

2.3 Time-dependent models

The time-dependent mechanisms are usually investigated by developing analytical methods adopting rheological models (comprising mechanical analogues) and empirical models based on laboratory testing data. Specifically, creep behavior is mathematically represented by the Burgers model. This model combines two simplified linear visco-elastic mechanical analogues in series: the Kelvin and the Maxwell that simulate a delayed manifestation of a static response due to boundary conditions alteration and a continued strain rate relaxation overtime under static boundary conditions, respectively shown in **Figure 7**.

Deformation that occurs at constant loading condition through time can be expressed using Eq. (1) [38], where: ε_1 is the axial strain, σ_1 is the constant axial stress, K is the bulk modulus, η_K is Kelvin's model viscosity, η_M is Maxwell's model viscosity, G_K is Kelvin's shear modulus, G_M is Maxwell's shear modulus. η_K, η_M, G_K, G_M are the visco-elastic parameters and are considered properties of the rock.

$$\varepsilon_1(t)=\frac{2\sigma_1}{9K}+\frac{\sigma_1}{3G_M}+\frac{\sigma_1}{3G_K}-\frac{\sigma_1}{3G_K}e^{-\left(\frac{G_K}{\eta_K}t\right)} \tag{1}$$

During stress relaxation, the strain-state is controlled and remains constant, thus rearranging Eq. (1) for a constant strain component, the material's stress state is changing according to Eq. (2).

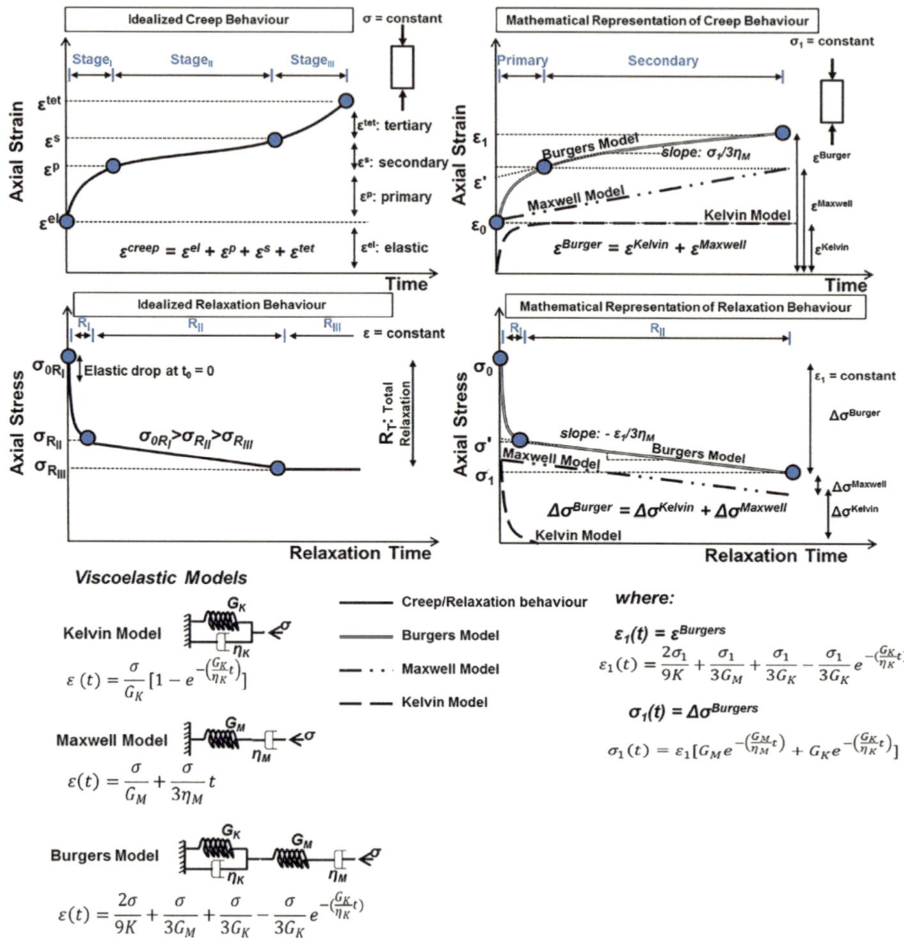

Figure 7.
Idealized creep and relaxation behavioral curves and the equivalent visco-elastic components in the Burgers model.

$$\sigma_1(t)=\varepsilon_1\left[G_M e^{-\left(\frac{G_M}{\eta_M}t\right)}+G_K e^{-\left(\frac{G_K}{\eta_K}t\right)}\right] \tag{2}$$

Goodman's [38] approach is usually adopted to derive the Bugers model parameters by curve fitting laboratory creep testing results. Using a similar approach for determining parameters and assuming that the material's behavior can be represented by the linear visco-elastic Burgers body in unconfined compression [33] found that the same parameters (i.e. viscosities and shear moduli) can be also derived from stress relaxation tests, (**Figure 7**).

In reality and embedded in this mathematical concept are the three stages of creep that follow the instantaneous response (0th stage) to changed boundary conditions resulting to a constant stress-state as follows:

- 1st stage or primary or transient creep where the delayed adjustment to a new equilibrium state takes place through visco-elastic (reversible) deformation, and may be accompanied by some irreversible behavior, resulting in strain accumulation with decreasing rate over time. This stage is commonly simulated with the Kelvin model analogue.

- 2nd stage or secondary creep where the material exhibits a consistent strain accumulation rate over time accompanied by inelastic distortion. The duration or even existence of this stage can vary depending on the ability of the rock type to transition from ductile to more brittle materials. The Maxwell visco-elastic model is commonly used to phenomenologically represent this stage.

- 3rd stage or tertiary creep where strong non-linear or accelerating strains occur (typically driving the material to rupture) due to strain-driven weakening, chemically related strength degradation and/or interaction of growing cracks. Visco-plastic models and/or so-called stress corrosion models are used to simulate tertiary creep.

A combination of Kelvin and Maxwell model components is referred to as the Burgers model which can be used to simulate stages 1 and 2 in combination.

2.4 Damage evolution and failure in brittle rocks

Over (geological) time, ductile deformation processes involve continuum mechanisms such as dislocation slip or migration of atomic vacancies within crystals resulting in distortion (pure or simple shear strain) [39]. However, in brittle materials, failure is controlled and governed by progressive damage driven by the pre-existing and new cracks initiation and evolution in the maximum load direction [40, 41]. **Figure 8** presents the four distinct stages during brittle deformability and failure: (i) closure of pre-existing cracks; (ii) linear elastic behavior; (iii) stable crack growth; and (iv) unstable crack growth, which leads to failure and the peak strength.

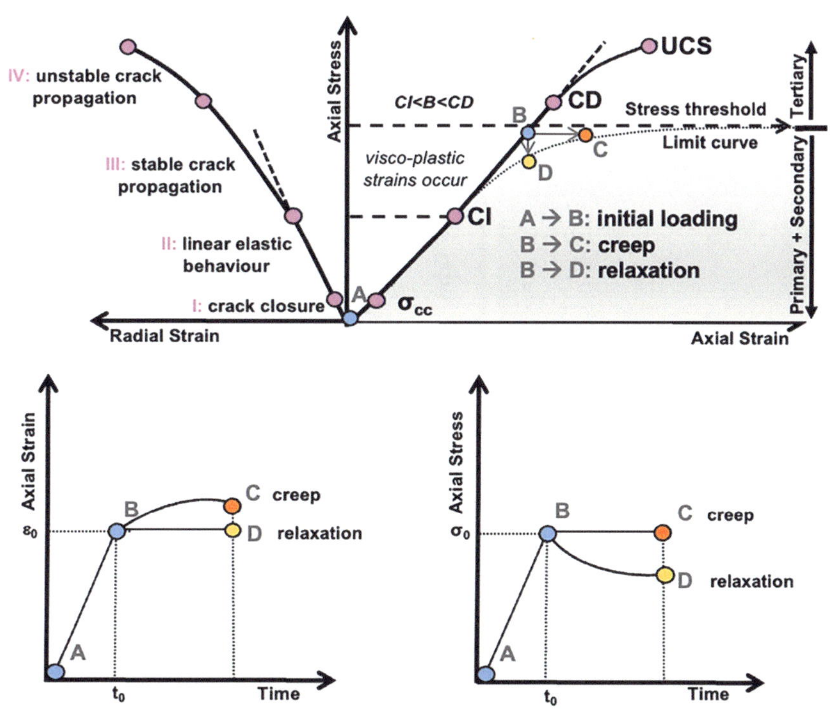

Figure 8.
Stress - strain response of brittle rock deformability and time-dependent behavior of creep and/or relaxation.

Stress–strain curves for brittle rocks can be used to determine the: (i) crack initiation stress (CI); (ii) critical damage stress or axial yield stress (CD), and (iii) uniaxial compressive strength (UCS). While UCS strength can inhibit the loading rate and testing procedure influences, CD is the true upper bound yield strength when obtained in the lab, according to ISRM [42] standards [43]. In the limit CD, can drop in situ to the lower bound defined by CI. This lower bound is relatively insensitive to moderate pre-existing damage and other influences and is found to be 30–50% of standard UCS in brittle rocks as measured in the lab [44] or by in situ back analysis [45]. Below CI, the sample is genuinely elastic, with no new damage occurring in the sample.

3. Laboratory testing program and methods

Laboratory tests were performed in two types of limestone to examine time-effects in brittle materials. The selected Jurassic limestone comes from a quarry north of Zurich, Switzerland (**Figure 9a**). The Cobourg limestone (**Figure 9b**) comes from the Bowmanville quarry near Bowmanville, Ontario, Canada. It should be noted that sample preparation were conducted according to ISRM [42].

Unconfined Compressive Strength (UCS) tests were conducted on 10 cylindrical samples of Jurassic and 9 of Cobourg. Relaxation tests were conducted on 19 Jurassic and 16 Cobourg samples. Static Load tests were performed on 12 Jurassic and 5 Cobourg samples.

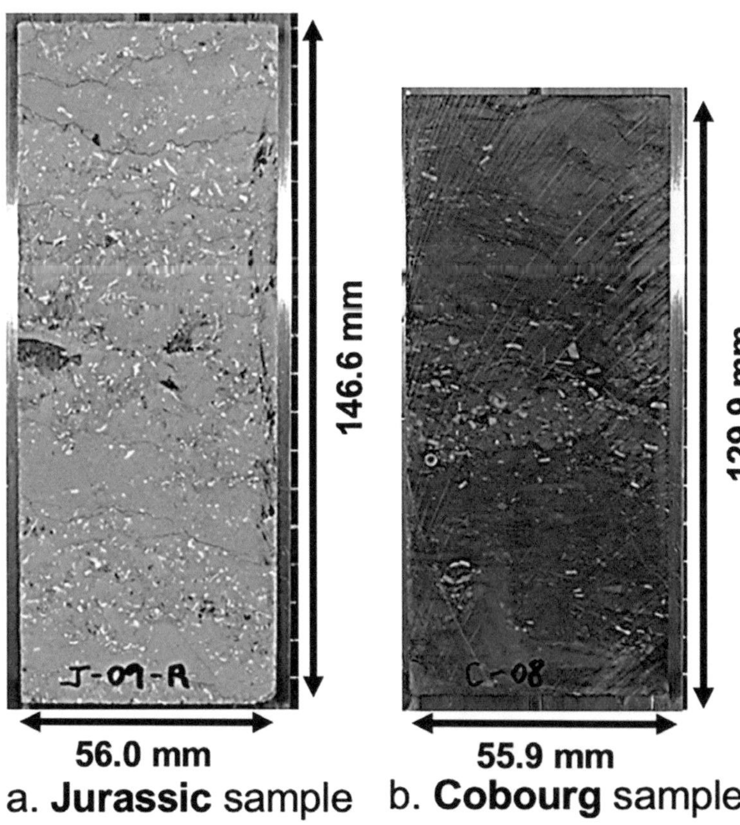

Figure 9.
Samples of (a) Jurassic limestone, (b) Cobourg limestone.

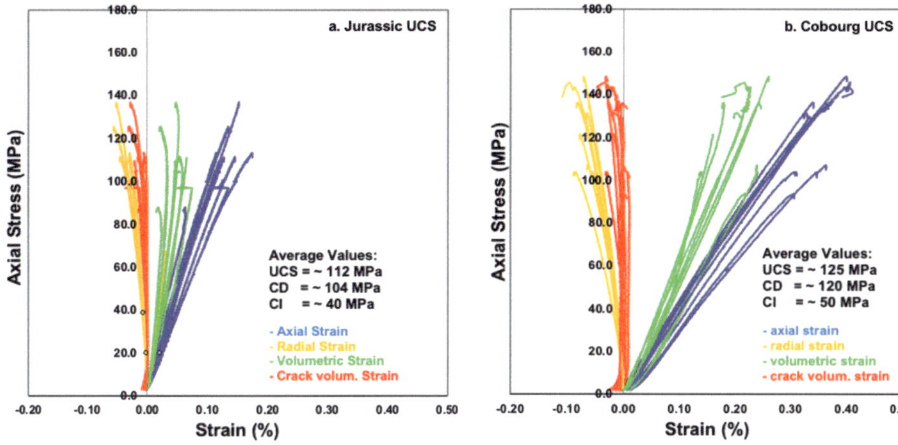

Figure 10.
Stress-strain response of limestone: (a) Jurassic samples and (b) Cobourg samples tested in unconfined compressive strength conditions.

3.1 Baseline testing series

The complete stress–strain curves of the UCS tests are shown in **Figure 10**. The average values estimated for UCS, CD and CI were 103 MPa, 91 MPa, 39 MPa, respectively for the Jurassic limestone and 125 MPa, 111 MPa and 50 MPa Cobourg limestone.

3.2 Relaxation testing series

Two test series have been performed: (i) Jurassic limestone was utilized to examine the applicability of various testing procedures (i.e. axial strain-controlled, radial strain-controlled, multi-step and single-step) for assessing the long-term relaxation behavior and (b) Cobourg limestone was performed utilizing a single-step axial strain-controlled testing procedure.

Emphasis was given to the maximum stress relaxation; the total change between the maximum stress value at the end of loading and before relaxation started. The lowest stress level resulted after relaxation with time. The steps of the procedure undertaken were the following:

- the maximum stress value before relaxation was recorded,
- the initial loading portion of the stress–strain curve was then removed,
- setting the time to zero at the point where the axial strain was kept constant,
- the load rate was kept the same for all the tests, and the initial loading duration ranged from 2 to 20 minutes,
- the axial stress was then normalized to the estimated average UCS,
- the maximum normalized stress was recorded and related to the maximum stress relaxation (the difference between the initial maximum stress and the minimum stress at the end of the relaxation test where no further relaxation took place).

The relations between the maximum stress relaxation and applied stress expressed as a driving ratio of UCS from all the relaxation testing series (axial strain-controlled) are summarized in **Figure 11**. It can be observed that there is an apparent trend between the multi-step and the single-step tests of Jurassic limestone. It can be easily seen that the multi-step tests exhibit less relaxation for similar driving stress-ratios than the single-step. The initial drop in stress mechanism that occurs rapidly for the first step of any test was attributed to being associated with the elastic energy within the sample and load system. A correction procedure was developed since the stress drop was associated with only the initial load stage. This stress drop was added to all subsequent load steps in the multi-step tests, shown (**Figure 11**) as corrected and exhibits similar amounts of stress relaxation compared with the single-step relaxation at a similar load level. Therefore, the multi-step tests, if corrected, can be conducted when limited samples are available [33]. Cobourg limestone shows a higher relaxation sensitivity as it exhibits more stress relaxation than the Jurassic limestone at the same stress levels.

3.2.1 Defining the three stages of stress relaxation

All the single-step test results showed a similar behavior during stress relaxation for both the limestones. This behavior can be characterized by three distinct stages, which were observed in the stress relaxation versus time graphs. An example of the test results is illustrated in **Figure 12**. The three stages can also be observed in the radial strain response with time, although there is a slight delay during the transition from stage to stage compared to the transition time of the stress relaxation shown as dt in **Figures 12** and **13**.

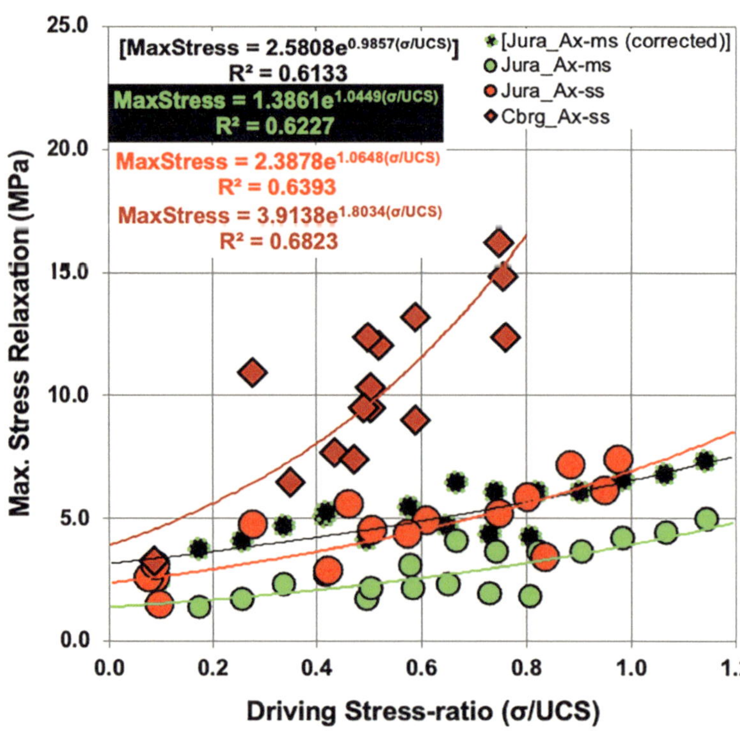

Figure 11.
Maximum stress–relaxation (MPa) to driving stress-ratio normalized to UCS of the single-step tests on the Jurassic and Cobourg samples, as well as the multi-step tests of the Jurassic samples. 'Ax' refers to axial strain-controlled conditions and 'ss' and 'ms' denotes single-step load and multi-step load tests, respectively.

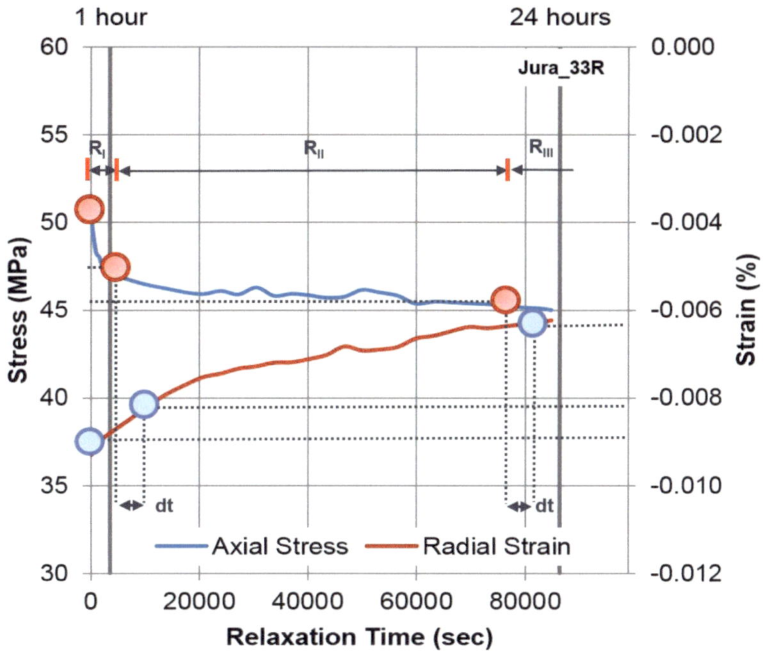

Figure 12.
The three stages of the stress relaxation process during a relaxation test under axial strain-controlled conditions illustrated on the Jura_33R sample.

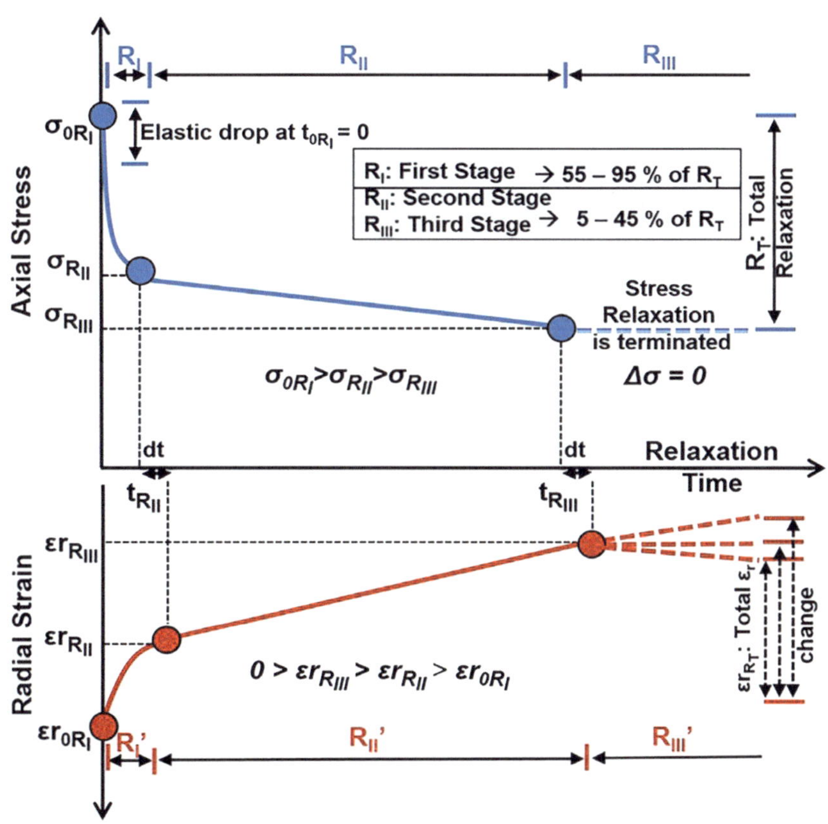

Figure 13.
The three stages of the stress relaxation process during a relaxation test under axial strain-controlled conditions.

When the axial deformation is kept constant, the stress relaxes at a decreasing rate; this period is defined as the first stage of stress relaxation (RI). At the end of this stage, the stress decrease approaches a constant rate, which marks the second stage transition (RII). The third stage of relaxation (RIII) follows where no further stress relaxation is measurable. At this stage, the stress reaches an asymptote, and the stress relaxation process is effectively complete, which others have observed [19]. Some samples did not exhibit the second stage of relaxation (RII), and in the first stage, 55% to 95% of the total stress relaxation takes place.

The radial strain does not always reach an asymptote. In this case the material is subject to a practically constant axial stress state with ongoing additional absolute radial strain decrease. This response is possibly related to a combination of three-dimensional visco-elastic response and crack behavior during stable propagation (in the axial direction) under constant axial strain.

The significance of this scientific observation should be considered during the excavation of an underground opening. Energy release and stress relaxation in such conditions commonly take place at the face of the excavated tunnel. The created free space disturbs the stress regime of the in-situ conditions. For the stress to re-distribute itself to a new equilibrium state, the rock mass tends to "relax" through the structural geological imperfections (i.e. discontinuities, fractures, joints) of the surrounding rock mass or the newly created fractures due to the excavation method and techniques used. In relation to the scientific observation of the three stages (**Figure 14**), it would be expected that the rock mass would relax in distinct but possibly overlapping stages. This can serve as an explanation of the sound of cracking closer to the tunnel face without observed failure. Another component of stress relaxation is the duration of this phenomenon until it is terminated. Knowing the duration of stress relaxation can be valuable in the support design and the installation timing, avoiding safety implications arising from support overstressing or resulting in cost savings.

3.3 Static load testing series

Single-step static load tests were conducted on 10 Jurassic, and 4 Cobourg samples and they were held at stress levels above CI for seconds to several days

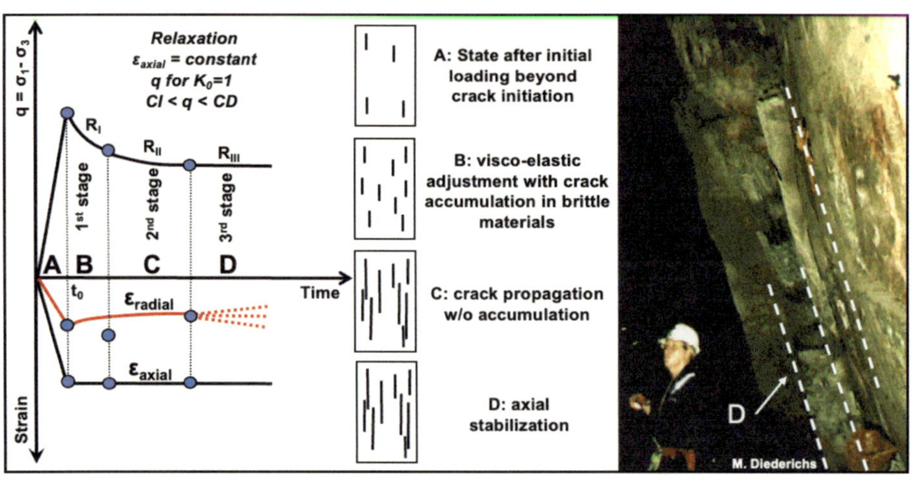

Figure 14.
The three stages of the stress relaxation process during a relaxation test under axial-strain–strain-controlled conditions and the response of the material during each stage, dashed lines on the right photo show axial stabilization of damage.

until failure occurred. Most of the single-step tests failed within the first few hours. Several samples did not fail after several days to weeks, at which time the test was terminated. While more practical and convenient, the single-step tests require the testing of more specimens to fully cover the spectrum of the expected range of time to failure. Multi-step tests were performed on three samples, 2 Jurassic and 1 Cobourg, to compare with stress levels derived from the single-step tests. The stress difference between the steps (varying between 2–4) was 5 MPa, and the duration of each step varied from 1 hour up to 10 days until failure took place. A few Jurassic samples did not fail, and it was decided to terminate these tests and unload the samples. To examine the long-term strength and time to failure of a material, the specimens need to fail under a constant load.

The static load testing began at load levels close to the peak strength, based on the Baseline Test results. Subsequent tests were conducted at lower driving stress levels approaching CD and below. In these tests, the constant target stress is applied and maintained by controlling the axial load while measuring the strains (axial and lateral) that increase as the sample proceeds toward failure. Samples loaded close to the peak strength fail catastrophically into many fragments, while samples loaded closer to CD fail less violently. Selected results are presented in this section, serving as examples to describe the main influencing factors during the two limestone tests' creep process.

Two aspects of time-dependency were examined: the first was to derive visco-elastic (creep) parameters for use in the Burgers model (or related models), and the second, the time to failure. Samples that did not fail were also examined to assess the potential reason why some samples fail, and others do not, even at the same driving stress-ratio. For this reason, this section focuses on analyzing and comparing the data from this testing series to other data available in the literature.

However, during the loading phase, the properties of the sample can be determined, such as the stiffness or the damage thresholds. The steps of the analysis procedure were:

- the maximum stress value at which the axial load was held constant was recorded.
- the initial loading portion of the stress–strain curve was used to estimate CI stress thresholds.
- the load rate was similar for all the tests and depending on the instantaneous stress level the initial loading duration ranged from 5 to 10 minutes, according to the ISRM [42] guidelines.
- setting the time to zero at the point where the axial load is kept constant,
- the maximum stress was normalized to an estimated UCS value for comparison to the literature.
- the maximum stress was normalized to the CI value from each sample test, as it is an independent value.
- the visco-elastic parameters were determined.

All the results presented refer to unconfined conditions.

3.3.1 Estimating the driving stress-ratio

In the literature, most of the testing results are presented in the form of time against the driving-stress-ratio, defined and used as the stress normalized by the strength of the sample. In most cases, the UCS is taken as an average value from standard UCS tests. In this section, a new solution is presented to examine similar datasets.

[45] suggested that there is a consistent relationship between UCS and CI for brittle rocks. The author has found this to be true for a number of test series with similar lithologies and compatible testing protocols [46]. It was decided to convert the CI values from this study's static load tests to an equivalent UCS value. The CI and UCS values from the Baseline testing series for the two types of limestone were used to develop the conversion factor (here: 2.66 for Jurassic and 2.52 for Cobourg), shown in **Figure 15**. The conversion factors were multiplied with the CI values estimated from the loading portion of the static load test for each sample. [26] suggested that the modified UCS* can be calculated using Eqs. (3) and (4):

$$UCS^{*} = a * CI \tag{3}$$

$$a = \frac{UCS^{B}}{CI^{B}} \tag{4}$$

where: UCS* is the estimated UCS, CI is the Crack Initiation value derived from the static load test, α is a constant and describes the slope of the CI versus UCS graph, and the superscript B denotes values from the Baseline Testing.

When the data (red circles and squares) are compared with other static load test results from various rock types, the time to failure of the samples from this

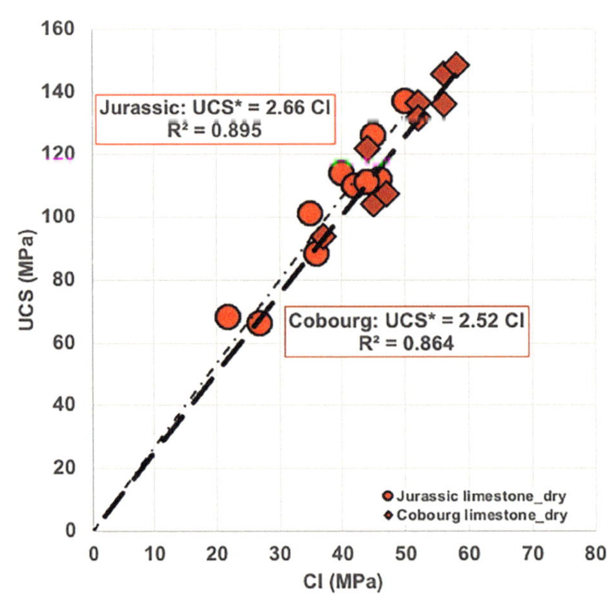

Figure 15.
The relationship between UCS and CI for the Jurassic and Cobourg limestone from the baseline testing.

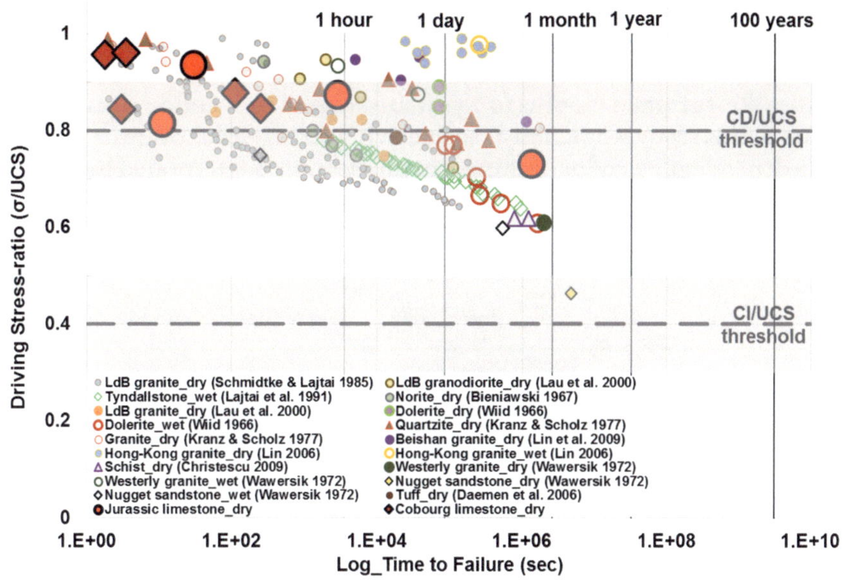

Figure 16.
Static load test data for hard rocks performed at room temperature in wet or dry conditions (where the driving stress-ratio is the stress level at failure to unconfined compressive strength of the material).

study seems to follow a similar trend (**Figure 16**). There are no samples loaded below the CI threshold that fail from the data presented and gathered from the literature.

Figure 17 categorizes the data according to the main rock type, sedimentary, metamorphic, and igneous.

The sedimentary rocks appear to follow a similar trend with the metamorphic rocks. In contrast, igneous rocks show more scatter because most test results have been on igneous rocks and that there are fewer results on sedimentary and metamorphic. There could also be due to different grain sizes of the granitic rocks tested characterized by grain-scale heterogeneity.

Granites and limestones, even though they fail similarly following brittle failure theory principles, their long-term strength is directly dependent on lithology, as better shown in **Figure 18**. Due to heterogeneous mineralogy and their different intrinsic properties, granitic rocks allow other creep behavior within different constituent crystal grains. Steady creep creates mechanical conflicts between the different grains and damage results. This creep-induced damage process is less dominant in monominerallic limestones, and therefore creep can occur with less resultant weakening.

Differences in the trend start to emerge when examining individual sample sets. The latter is partly because there is a lack of statistically representative data sets on an individual sample set, except the lac du bonnet set.

From **Figure 19**, it is evident that above 0.8 ucs or the cd threshold, all samples failed within an hour. Below the ci threshold, where pre-existing cracks are closing, and elastic strains govern, no failure should occur as [47, 48] reported from testing cobourg limestone samples for up to 100 days. Commonly, the static load stress levels fall between the ci and the cd thresholds. This region is an uncertain region since between ci and cd crack propagation, and accumulation of damage occurs in the short-term. Still, in the long-term, the time component can degrade the rock, further leading it to failure. However, below 0.7 ucs, no failure is shown. These no-failure points could be the result of not holding the load constant for long enough. Data from the literature suggests that failure could be expected at such driving stress-ratios. Tests from 6 months to 1 year are advised to examine if samples of the limestones in this study would fail at such driving stress-ratios over the long-term. The time-dependent behavior discussed in this section is interpreted to be, in part, the result of the behavior of new microcracks, the intensity of which impacts the final ucs value [49].

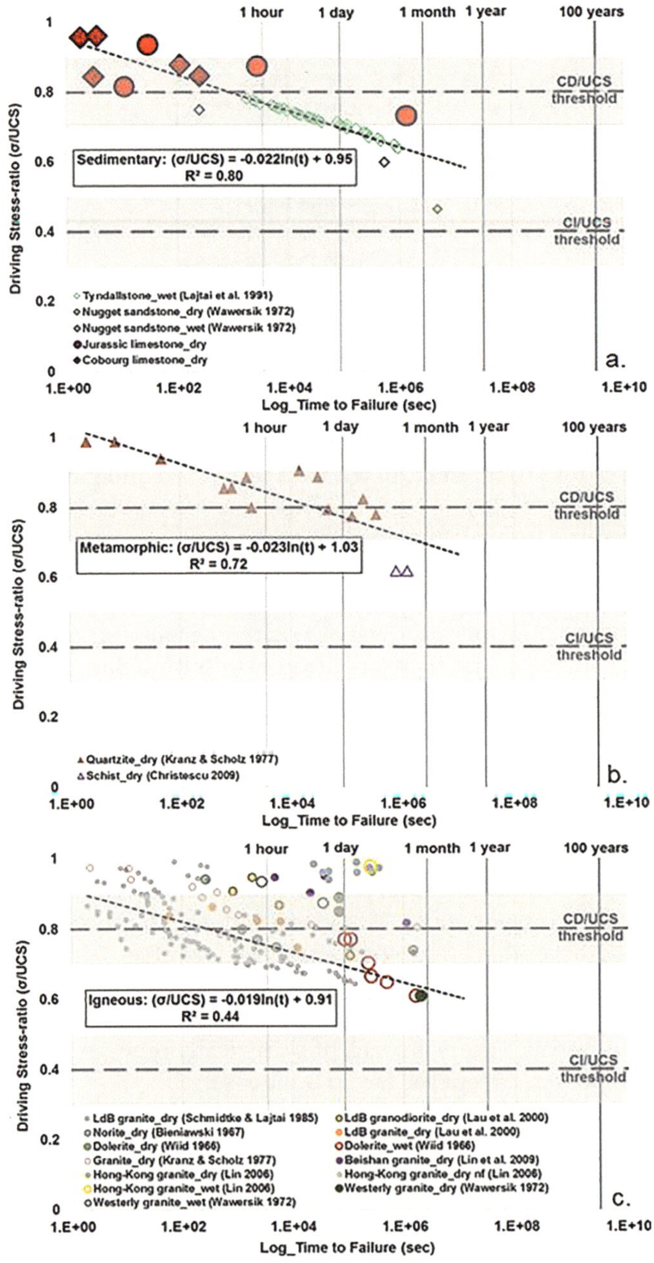

Figure 17.
Static load test data for: (a) sedimentary, (b) metamorphic, (c) igneous rocks performed at room temperature in wet or dry conditions (where the driving stress-ratio is the stress level at failure to unconfined compressive strength of the material).

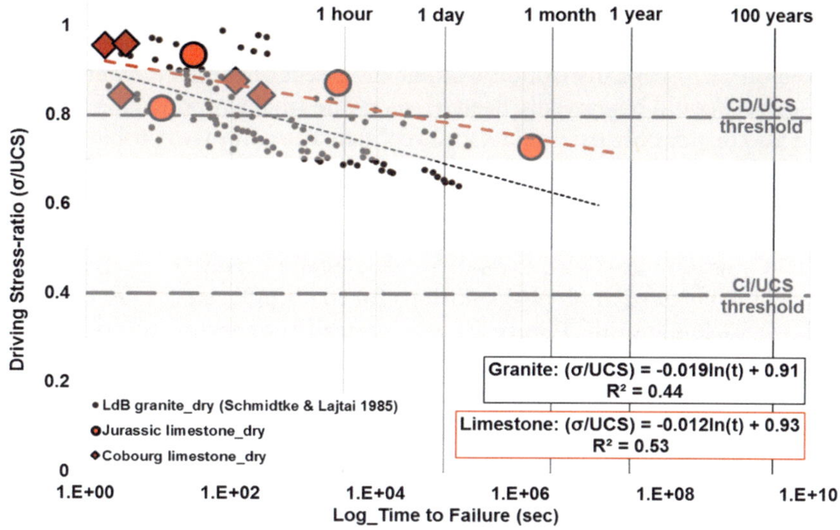

Figure 18.
comparison of static load test data on limestone and granite performed at room temperature in dry conditions (where the driving stress-ratio is the stress level at failure to unconfined compressive strength of the material).

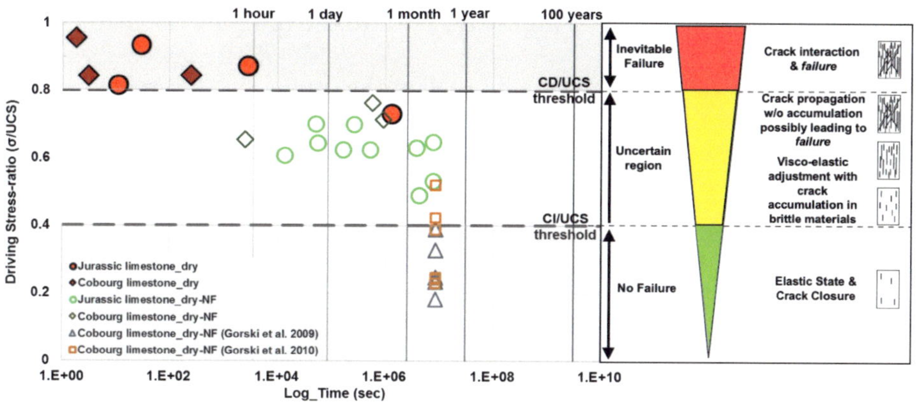

Figure 19.
static load test data of jurassic and cobourg limestone performed at room temperature in dry conditions (where the driving stress-ratio is the stress level at failure to unconfined compressive strength of the material). The 'nf' in the legend indicates samples or tests did not fail whereas the 'f' denotes samples or tests reach failure.

4. Time-dependent effects in tunneling

Time-dependent deformations associated with rock tunneling are a reality that warrants further investigation and understanding and can be observed during excavation and/or after the construction period of the project.

4.1 Analysis of time-dependent rock masses using the convergence-confinement method

Understanding the nature and origin of deformations due to an underground opening requires, as [50] noted, both knowledge of the rock-support interaction and field data interpretation. This tunnel wall movement, also known as convergence, results from both the tunnel face advancement and the time-dependent behavior of the rock mass.

The Convergence-Confinement Method (CCM) is a two-dimensional simplified approach that can be used to simulate three-dimensional problems. Analytical solutions based on CCM (usually examine either the effect of tunnel advancement or the time-effect) could be partially used to select the final support. One may wonder if it could also be possible to simulate and replicate the complete problem. Time-dependency is acting during the timeframe of construction impacts, the so-called Longitudinal Displacement Profiles (LDPs) for deformation estimation during tunnel advance. LDP is an accompanying tool used with the Ground Reaction Curves (GRC) used in the CCM to relate internal wall pressure relaxation to tunnel displacement. Suggested analytical solutions for LDPs [50], etc.) refer to elastic or elasto-plastic rock materials. **Figure 20** schematically illustrates the effect of both time and tunnel advancement on the LDP of a tunnel excavated in a visco-elastic medium. The tunnel's advance is simulated by reducing the internal pressure, pi, initially acting on the tunnel core (as p0). The rock responds by convergent deformations (via the GRC), which are, in turn, linked to the tunnel advance via the LDP. This aspect of time-dependency is also discussed, examined and further analyzed in this chapter.

Analytical and closed-form solutions that consider the time-dependent convergence have been proposed in the literature for supported and unsupported tunnels with linear and non-linear visco-elastic medium [51, 52], etc. Most of these formulations also consider the tunnel advance in the estimated total deformation yet are found to be impractical due to the complex calculations required [53, 54].

Figure 21 illustrates the anticipated LDP of the tunnel displacement in an elasto-visco-elastic medium where no tertiary creep takes place. More ductile materials, as in the case of rock salt, can behave in such a manner. It is shown that when no time-effect is considered, the total displacements are underestimated, which can lead to erroneous calculations at the initial stages of the design process.

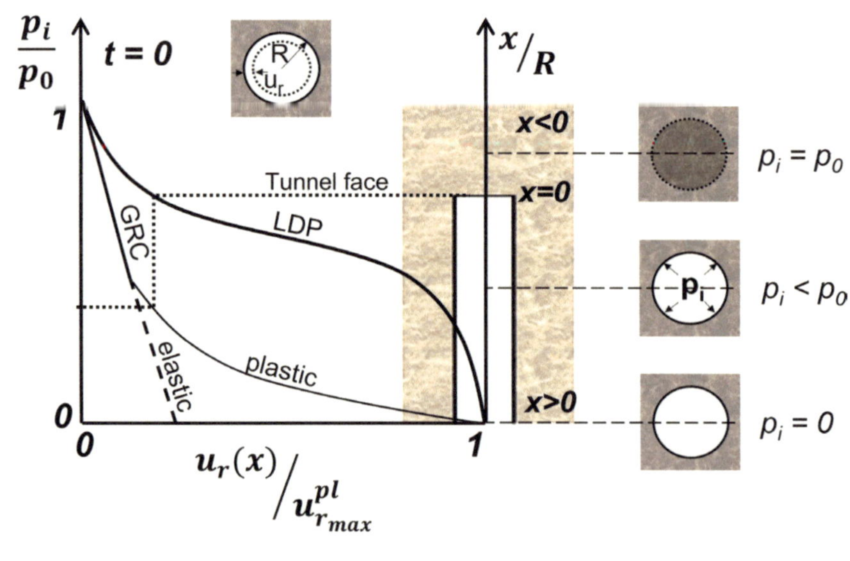

Figure 20.
Schematic representation of the GRC of an elastic (t = 0) and a visco-elastic material (t > 0) and their relation to the LDP. Y-axis on the left refers to the internal pressure (pi) normalized to the in-situ pressure (po), Y-axis on the right refers to the distance from the face (x) normalized to the tunnel radius (R) and X-axis refers to the radial displacement at a location x normalized to the maximum radial displacement due where t denotes time and subscripts e and ve refer to elastic and visco-elastic material, respectively.

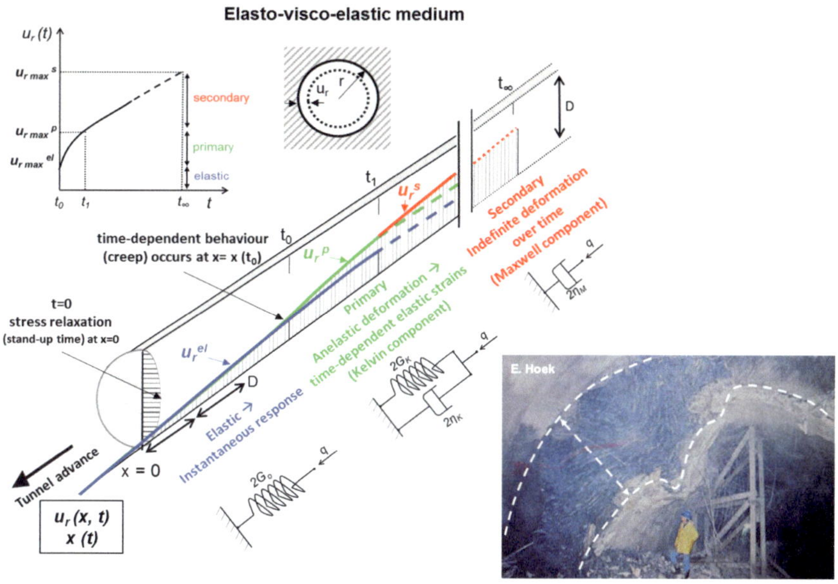

Figure 21.
Schematic representation of the longitudinal displacement profile (LDP) in an elasto-visco-elastic medium.

4.2 Capturing the time-effect in tunneling using a new numerical approach

An axisymmetric parametric analysis was performed within FLAC software. The geometry of the model and the excavation sequence characteristics are shown in **Figure 22**. A circular tunnel of 6 m diameter and 400 m length was excavated in isotropic conditions. Full-face excavation was adopted. Two cases of a drill and blast (Case 1: D&B) and a TBM (Case 2: TBM) were assumed depending on the excavation step 1 m and 3 m, respectively. The rock mass was set to behave as an elasto-visco-elastic material using the CVISC model. The analysis aimed to examine the contribution of primary and secondary creep. In this regard, the Maxwell body's viscous dashpots within the CVISC model were deactivated and reactivated,

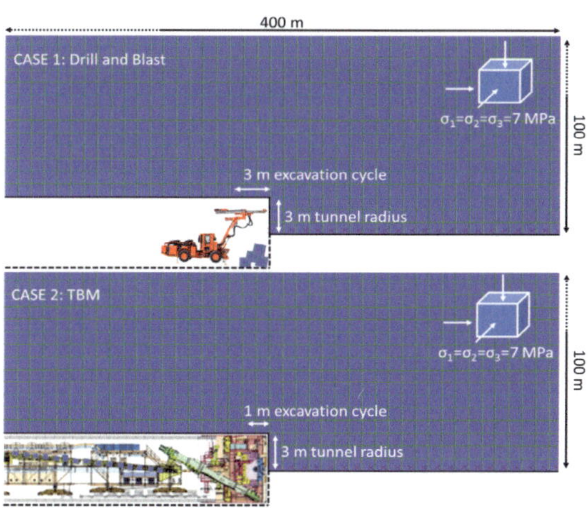

Parameter	SET #1	SET #2
	Rockmass conditions	
γ (KN/m³)	20	20
φ (°)	23	23
c (MPa)	1.00E+20	1.00E+20
t (MPa)	1.00E+20	1.00E+20
K (MPa)	1.23E+03	4.32E+04
	Stress Conditions	
k_0	1	1
p_0 (or σ_0) (MPa)	7	7
	Visco-elastic Parameters*	
G_K (MPa)	4.98E+02	6.73E+04
η_K(MPa*s)	1.34E+08	1.72E+09
G_0 or G_M (MPa)	5.66E+02	1.99E+04
η_M (MPa*s)	8.82E+08	1.20E+10
$T_K = \eta_K/G_K$ (s)	2.69E+05	2.56E+04
Reference*	Barla et al. 2010	Zhang et al. 2010

Figure 22.
(Left) Schematic illustration of the excavation sequence used within the numerical axisymmetric analysis; case 1 refers to drill and blast method with 3 m excavation step per cycle; case 2 refers to TBM (Tunneling boring machine) method with 1 m excavation per cycle. (Right) Parameters used for CVISC model.

accordingly. Besides, two different sets of visco-elastic (creep) parameters were used then for both cases shown in **Figure 22**. Furthermore, the excavation cycle duration was also simulated to consider both the time-dependent component and the tunnel advance representing the real conditions in a tunnel problem and varied from 2 to 8 hours. In addition, two supplementary analyses were performed: with the Kelvin-Voigt model and with the elastic model to validate the numerical models and compared with analytical solutions.

It should be stated that the visco-elastic parameters were chosen according to the analytical solution (Eq. (5)) of the Kelvin-Voigt model developed by [51].

$$u_r = \frac{\sigma_o r}{2G_o} + \frac{\acute{o}_o r}{2G_K}\left[1 - \exp\left(-\frac{t}{T_K}\right)\right] \tag{5}$$

(where: σ_0 is the in-situ stress conditions, r is the tunnel radius, G_0 the elastic shear modulus, G_K is the Kelvin shear Modulus, η_K is Kelvin's viscosity and T_K is known as retardation time and it is the ratio of Kelvin's viscosity over the Kelvin Shear Modulus and is indicator of when the model will convergence and reach a constant value.)

The selected retardation time (T_K) varies one order of magnitude between the two sets as it controls the curvature of Kelvin's model behavior. The following **Figures 24** and **25** 'x' is the distance from the tunnel face, R is the tunnel radius, ur is the absolute radial tunnel wall displacement, $ur_e max$ is the maximum elastic displacement and $ur_\infty max$ is the maximum visco-elastic displacement of the Kelvin-Voigt model. Gray and black lines are the elastic and the zero-viscosity KV models respectively.

4.2.1 Primary stage of creep, KELVIN-VOIGT (KV)

The Kelvin-Voigt model was assumed to represent the primary stage of creep and was used to simulate an elasto-visco-elastic rock mass's mechanical behavior. The results for both cases are presented in **Figure 23**. They imply that increased cycle time or excavation delay exacerbates the rock mass's mechanical behavior; as in all models, an increase of the ultimate total displacement was observed. This increase depends on the visco-elastic parameters of the Kelvin-Voigt model. The increase of the retardation time will increase the time required by the model to reach a constant value and become time-independent.

The deviatoric stress was related to the displacement data normalized to the maximum displacement of the Kelvin-Voigt model ($ur_\infty max$). Time-dependent behavior starts for both cases when the deviatoric stress reaches a critical value (q_{cr}) shown in **Figure 23b**. This critical value is attained after one excavation step at the point which the time-dependent LDPs deviate from the elastic LDP. In the drill and blast case, this is 3 m away from the tunnel, whereas for the TBM case it is 1 m.

4.2.2 Secondary stage of creep, Burgers (B)

The second stage of this analysis was to investigate the influence of both primary and secondary creep behavior stages using the Burgers model. The results presented in **Figure 24** show the maximum strains due to the secondary stage (Maxwell) are effectively infinite. This is also observed on **Figure 21**. In this part, it was noticed that the magnitude of the total displacements between the two cases varied significantly. The excavation method influences the accumulated displacements.

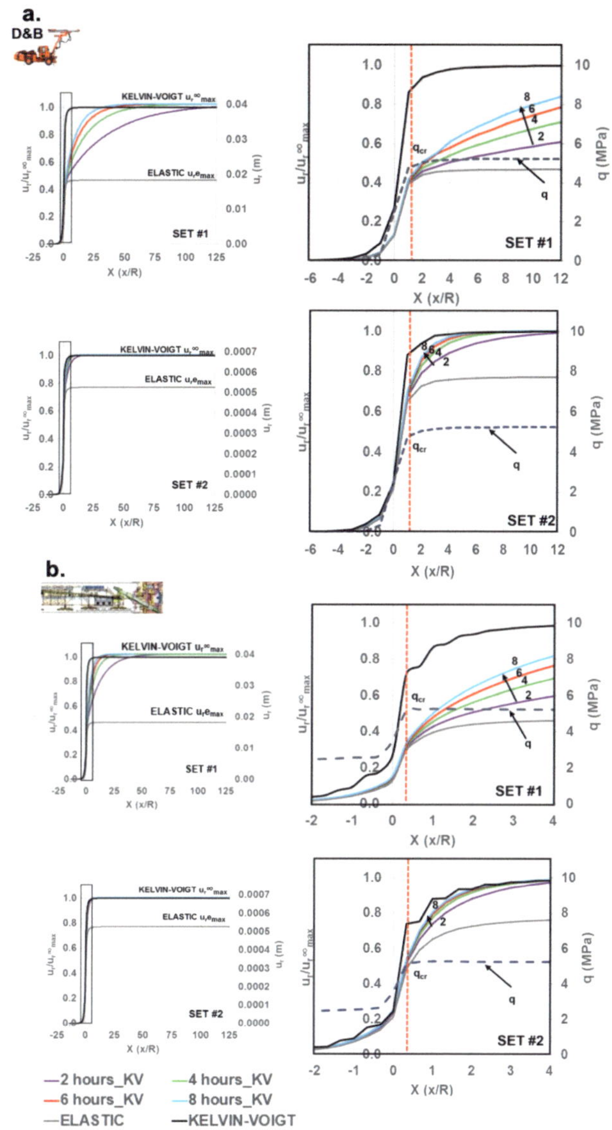

Figure 23.
(Left) Numerical results of LDPs, (right) closer representation of the data; relating the deviatoric stress (q) to the tunnel wall displacement normalized to the maximum displacement of the KELVIN-VOIGT model (ur∞max) for: (a) the drill and blast case (DB) and (b) the TBM case.

In the drill and blast case, all two sets of parameters exhibited less displacement than the TBM case for the same duration of the excavation cycles. During a TBM tunnel excavation, the tunnel excavation requires more time than a drill and blast excavation for the same excavation cycle. For instance, a TBM that excavates 1 m every 6 hours, the elapsed time is three times longer than the drill and blast case of 3 m excavation per cycle. In the TBM case, the time for the excavation of the same length tunnel will result in an accumulation of displacement increase. However, this may not always represent real conditions as TBMs are commonly preferable since they tend to achieve better excavation rates; if proven affordable. Suppose the latter is the case, then a TBM excavation of a two-hour excavation cycle. In that case, it is shown that the surrounding rock mass represented by SET#1 exhibits less displacement than an eight-hour excavation cycle using drill and blast.

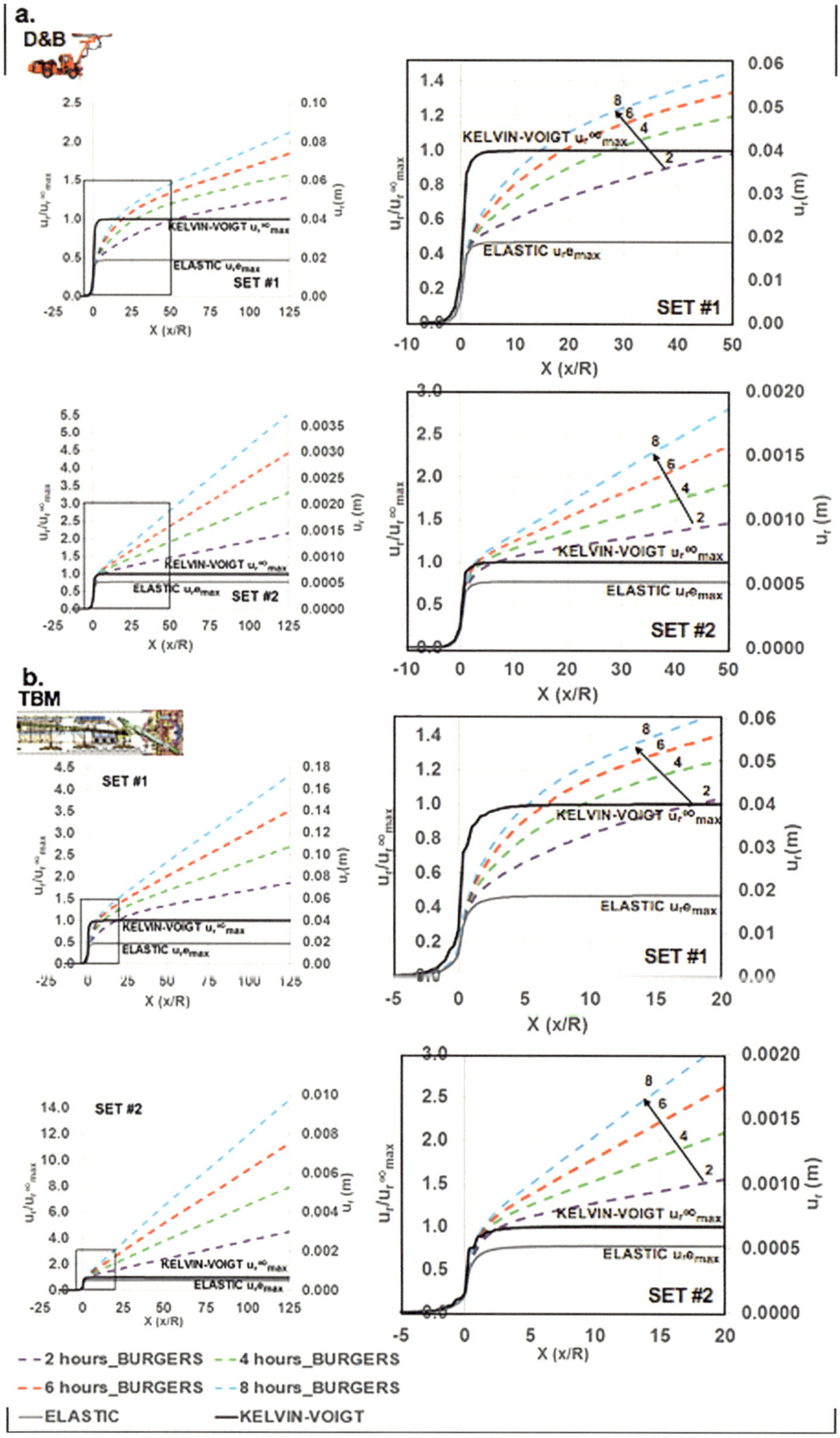

Figure 24.
(Left) Numerical results of LDPs for: (a) the drill and blast (DB) and (b) the TBM case of the BURGERS (B) analysis (the hours on the legend denote hours per excavation cycle), (right) closer representation of the data.

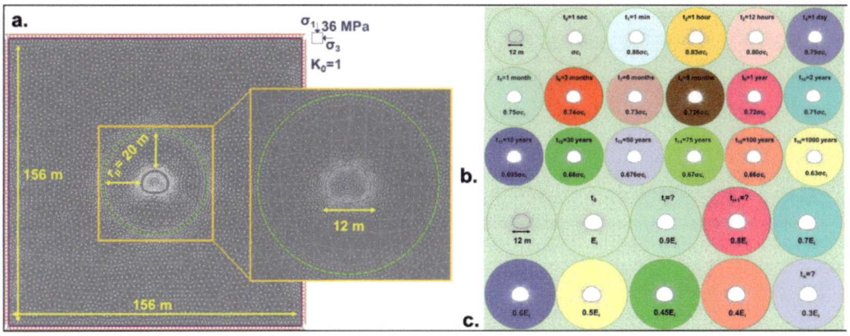

Figure 25.
(a) Geometry and mesh conditions of the model used, rp denotes the radius of the plastic zone, and incremental reduction of (b) intact rock strength according to long-term strength and (c) Young's modulus.

4.3 Predicting the long-term behavior of rock masses in tunneling

Although there are time-dependent models available to predict rock materials' rheological behavior potential, it is commonly observed that in two-dimension (2D) modeling, time-dependent behavior is not directly simulated using selected 2D coded software with noted limitations [1]. It is, therefore, necessary a method to be developed and 'pseudo' simulate this type of behavior. For instance, the plastic zone can be used as an indicator of the overall time-dependent displacements and calibrated to in-situ measurements or laboratory testing [1].

[25] proposed a new methodology for predicting rock masses' long-term behavior using the information derived when testing rock materials under constant loading, which results in strength degradation by 'pseudo-simulating' numerically this behavior. [25] examined two main sets of numerical models in plane strain conditions in-plane RS2 (Rocscience). The models' main difference was that the material included in the plastic zone changed parameters with time-steps in the one set of models. The first aimed at pseudo-simulating time-dependent behavior by using the Long-Term Strength (LST) according to strength-degradation Eq. (6) of the limestone bases on the laboratory data previously shown in **Figure 18**.

$$(\sigma/UCS) = -0.022\ln(t) + 0.95 \tag{6}$$

In addition, 19 stages were simulated, as shown in **Figure 25a**. Where in each stage, a new σci (strength of intact rock) was assigned only to the material of the plastic zone, according to Eq. (6). Each strength reduction represented a specific time from 1 second to 1000 years. The second set of analyses were based on Young's modulus (Ei) reduction from the initial 40 GPa to 12 MPa of 10% in every modeling stage (**Figure 25b**). It should be highlighted that the decrease in both strength and Young's Modulus reduction was applied to the plastic zone, assuming that the rest model behaves as an elastic material.

Every increment on the strength-degradation models was related to a time according to the lab results and time to failure graph shown in **Figure 18**, such as the YMR models can be associated with a specific time. For instance, a reduction of the intact strength of 21% (0.79 σci) can simulate the deformation acquired in 1 day and reflects the deformation of the 30% reduction of Young's Modulus (**Figure 26**). Moreover, to simulate the rock mass's deformation around the tunnel after a 2-year period, one can either reduce the intact strength to 0.71 σci or reduce the Young's Modulus to 50% (**Figure 26**).

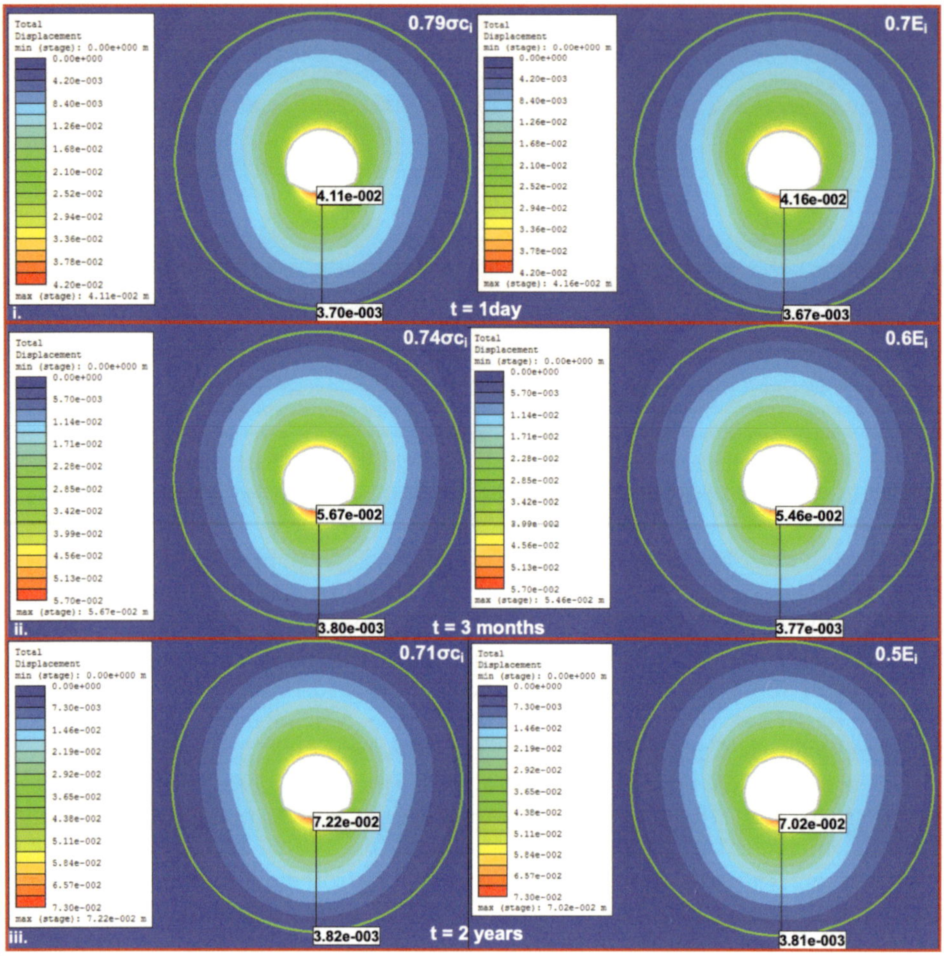

Figure 26.
Numerical resluts of total displacements of LST models (left colummn) and YMR (right colummn).

Relating the strength-degradation (or the LTS) with the YMR to time for specific lithologies can produce a database that one can use to capture the time effect on the rock mass behavior, as shown in **Figure 27a**. The yellow triangles reflect the YMR with the time, whereas the green circles the limestone's overall behavior based on the laboratory data (blue and light blue diamonds and squares). When the YMR method is used, the reduction factor can be estimated using Eq. (7), where t is time, and E/Ei is Young's Modulus-ratio.

$$YMS \text{ of limestone} : (\mathrm{E} / Ei) = -0.028 \ln(\mathrm{t}) + 1.014 \tag{7}$$

Using this YMR approach proposed by [25] for the granite (**Figure 27b**) the estimated reduction of the Young's Modulus is given in Eq. (8).

$$YMS \text{ of granite} : (\mathrm{E} / Ei) = -0.03 \ln(\mathrm{t}) + 1.018 \tag{8}$$

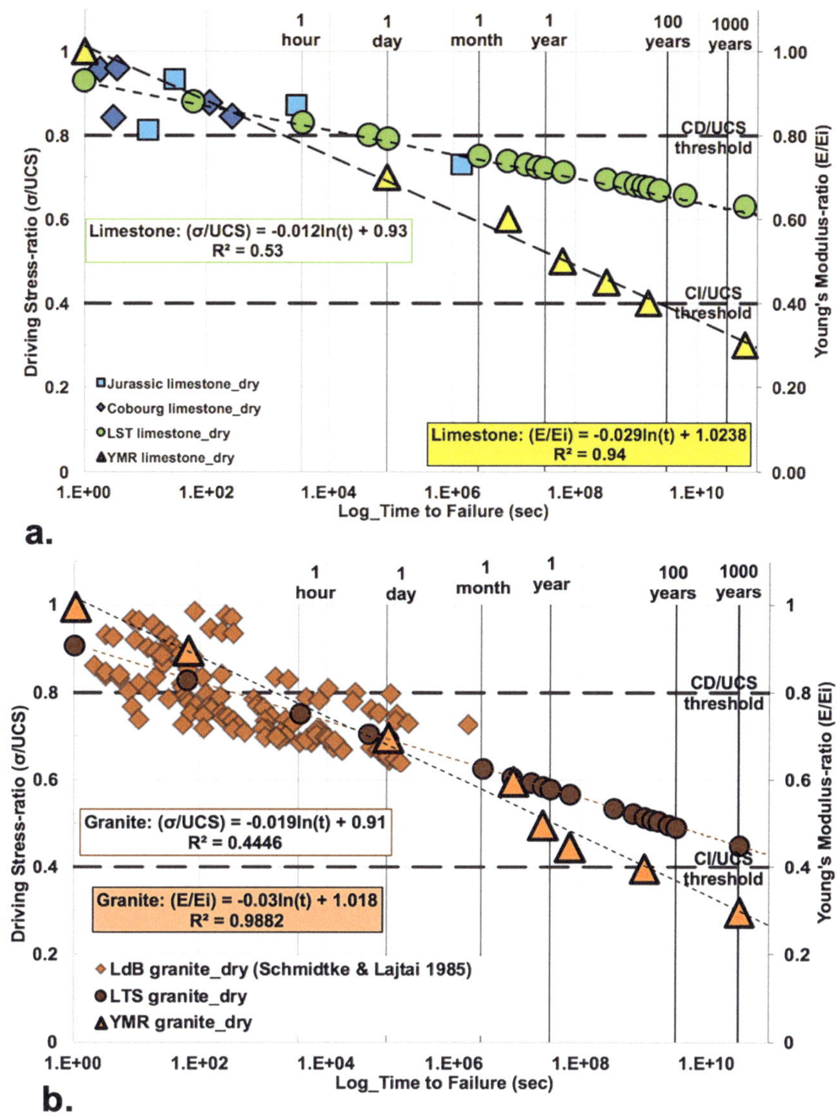

Figure 27.
Driving stress-ratio and Young's modulus-ratio in relation to time to failure from static load test data performed at room temperature in dry conditions, (where the driving stress-ratio is the stress level at failure to unconfined compressive strength of the material), for a. limestone and b. granite.

It should be stated that the analyses presented herein can be used for values of at least 0.5σci and higher as the below this threshold (CI), no failure is anticipated, below this threshold, the observed behavior is considered to be linear elastic.

5. Conclusions

This research work and the resultant publications presented in this Chapter have contributed to a better understanding of the "Time-dependent behaviour of rock materials". This section summarizes the key findings of this study.

It is widely accepted that significant research contributes to studying the time-dependent behavior of geo-materials and their effects in geoengineering applications by developing models. However, these models mainly focus on simulating the visco-elastic creep behavior and are developed based on the back-analysis of existing datasets. In many cases, these models can replicate sufficiently creep behavior during the primary and secondary stage when appropriate parameters are derived and used. Furthermore, the applicability of such models is commonly broader when dealing with weak rock masses. As a result, there is a knowledge gap when dealing with time-dependent behavior in brittle rock materials. Its effect is considered limited and usually neglected and covered by other progressive damage mechanisms. Nevertheless, this study has shown the importance of taking into consideration brittle time-dependent behavior, and it is recommended that engineers, scientists and practitioners utilize the existing models to simulate time-dependent behavior with appropriate parameters as with a few modifications, these models can capture the behavioral trend as long as the appropriate parameters are utilized.

In this study relaxation, the decrease of applied load at constant deformation was investigated and re-defined. It was shown that relaxation could be considered as an inversion of a creep behavior. It was concluded herein that axial strain-controlled tests are less sensitive to testing challenges during a relaxation test than a creep test. Single-step and multi-step tests have been performed in this study. It was shown that single-step is easier to perform, but there is sample consuming in order to obtain a complete dataset to cover the total stress spectrum. From the results, it was also shown that relaxation takes places when cracks initiate and propagate during the sample. It attains a constant value (asymptote) when axial crack stabilization is reached. The most important outcome of this work was identifying the existence of three distinct stages that occur during time-dependent stress relaxation. These three stages (R_I, R_{II}, and R_{III}) were introduced and clearly defined. The first two stages are similar to the first two stages in creep behavior. In contrast, the third stage differs as the sample reaches a stable condition compared to tertiary creep where it reaches failure.

Another set of tests, static load, both single-step and multi-step, are presented here. This time the axial load (stress) was kept constant, focusing on time-dependent behavior over time. Once again, it is shown that multi-step tests might be advantageous in terms of deriving visco-elastic parameters in different target stress levels using only one specimen; however, when considering time-to-failure single-step tests are preferred. In this section, two types of limestone (Jurassic and Cobourg) were investigated, and the time-to-failure behavior was compared to other rock types from data published in the literature. It was concluded that an overall trend does exist. This general trend was scrutinized at a second stage based on rock types providing specific trends for sedimentary, metamorphic, igneous, which can be used to predict time-to-failure for laboratory samples:

- Sedimentary: $(\sigma/UCS) = -0.022\ln(t) + 0.95$
- Metamorphic: $(\sigma/UCS) = -0.023\ln(t) + 1.03$
- Igneous: $(\sigma/UCS) = -0.019\ln(t) + 0.91$

where σ/UCS is the driving stress-ration and t refers to time.

The limestone dataset was compared to the widely used dataset of Lac du Bonnet granite, showing that the limestone's long-term strength is higher than the granite. The latter means that the limestone can withstand longer time-depended behavior than the granite. The latter can be explained by the fact that these two rock types differ in their mineralogical structures. The granite's increased heretegoeneity contributes to different creep rates of the various grains (quartz, feldspars, mica). The latter generates incompatible strains over time, causing micro-cracking. Similar processes do occur within the limestone, but due to its homogeneity, creep is constrained in calcite (monomineralic) and is associated with less damage increasing the time up to failure. It was also observed that both limestones that failed at a stress threshold above 0.8 UCS failed within the first 60 minutes. On the contrary, below CI threshold, no failure was observed, and between 0.5 to 0.8 UCS, failure will take place at some point between the first hours to months, depending on the rock type. Another outcome of this work was the identification of Maxwell's viscosity threshold as an indicator of failure. This observation can explain why some specimens fail and some others did not (yet).

Time-dependent behavior during tunneling can play an important role in the project success in the design and, most notably, in the construction process. This fourth dimension (time-effect) in tunneling was investigated numerically by performing an axisymmetric parametric analysis. From the research was concluded that current conventional methods adopted to predict the Longitudinal Displacement Profile of tunnel displacements have limited applications and fail to capture the overall displacement over time. It was also presented that both the excavation methods and excavation rate (tunnel advancement rate) can affect (deteriorate) the mechanical behavior of the surrounding rock mass. In this work, only creep behavior was considered a contributor to time-dependent deformation and was simulated with the modified purely visco-elastic CVISC model assuming the rock mass as a visco-elastic medium. It was further concluded that the retardation time (in the Kelvin-Voigt model) does control the timing at which the maximum tunnel displacement is reached during the primary stage of creep.

Finally, a new but yet simple tool that can be used to predict the long-term behavior of brittle materials as limestone using either the Long-Term Strength (LTS) approach (strength-degradation) and the Young's Modulus-Ratio (YMR) was presented. It should be stated that the proposed methodology should be used as a first estimate to relate the strength-deterioration of the rock material over time. Furthermore, input parameters can be derived in the plastic zone around an underground opening using this approach that can then be used in numerical analyses similar to the one presented herein.

Author details

Chrysothemis Paraskevopoulou
School of Earth and Environment, University of Leeds, Leeds, United Kingdom

*Address all correspondence to: c.paraskevopoulou@leeds.ac.uk

References

[1] Paraskevopoulou, C., Vlachopoulos, N., Diederichs, M.S. 2012. Long-term behaviour and support analysis using numerical modeling, In: Proceedings of the 46th US Rock Mechanics – Geomechanics Symposium ARMA. June 2012, Chicago, USA.

[2] Paraskevopoulou, C., Oke, J., Vlachopoulos, N. 2012. Practical modelling approaches to determine the long term behaviour of tunnel construction. In: Proceedings of the Canadian Tunnelling Symposium TAC on Tunnels and Underground Spaces: Sustainability and Innovations. September 2012, Montreal, Canada.

[3] Paraskevopoulou, C., Diederichs, M.S. 2013. A comparison of viscous models under constant strain and constant stress: Implications for tunnel analysis. In: Proceedings of the World Tunnel Congress on Underground – the way to the future. June 2013, Geneva, Switzerland.

[4] Paraskevopoulou, C., Diederichs, M.S. 2013. A comparison of viscous material model mechanics and impacts under different boundary conditions for tunnels and shafts. In: Proceedings of the 47th US Rock Mechanics – Geomechanics Symposium ARMA, June 2013, San Francisco, USA.

[5] Paraskevopoulou, C., Benardos, A., 2012. Construction cost estimation for Greek road tunnels in relation to the geotechnical conditions, In: Proc. Int. Symp. Practices and Trends for Financing and Contracting Tunnels and Underground Works (Tunnel Contracts 2012), March 2012, Athens.

[6] Paraskevopoulou, C., Benardos, A. 2013. Assessing the construction cost of tunnel projects. Tunnelling and Underground Space Technology Journal, 38(2013), 497-505. https://doi.org/10.1016/j.tust.2013.08.005

[7] Benardos, A., Paraskevopoulou C., Diederichs, M., 2013. Assessing and benchmarking the construction cost of tunnels. In: Proceedings of 66th Canadian Geotechnical Conference, GeoMontreal on Geoscience for Sustainability, September 29–October 3, 2013, Montreal, Canada.

[8] Paraskevopoulou C., Boutsis, G., Cost Overruns in Tunnelling Projects: Investigating the Impact of Geological and Geotechnical Uncertainty Using Case Studies. Special Issue Underground Infrastructure Engineering of Infrastructure Journal (MDPI). Infrastructures 2020, 5(9), 73; https://doi.org/10.3390/infrastructures5090073

[9] Paraskevopoulou, C. 2016. Time-dependency of rock and implications associated with tunnelling, PhD Thesis. In: Queen's University Publications. Canada.

[10] Weaver, S.H. 1936. Creep curve of steel. Trans Amer Soc. Mech. Eng., LVIII, 745-751.

[11] Griggs, D. 1939. Creep of Rocks. Journal of Geology. 47, 225-251

[12] Bieniawski, Z.T. 1974. Geomechanics classification of rockmasses and its application in tunnelling. In: Proc. of the 3rd Congress, ISRM, Denver.

[13] Barton, N.R., 1974. Estimating the shear strength of rock joints. In: Proc. of the 3rd Congress, ISRM. Denver.

[14] Palmstrom, A. 1995. RMi-A rockmass characterization system for rock engineering purposes. Ph.D. Thesis. University of Oslo, Norway, 400 pages.

[15] Widd, B.L. 1966. The time-dependent behaviour of rock; Considerations with regard to a research program. CSIR Report MEG 514,

Rock Mechanics Division, National Mechanical Engineering Research Institute, Pretoria, South Africa.

[16] Bieniawski, Z.T. 1967. Mechanism of Brittle Fracture of Rock, parts I, II, and III. Rock Mech. Min Sci, 4 (4), 395-430.

[17] Wawesik, W.R. 1972. Time-dependent rock behaviour in uniaxial compression. In: Proc. of the 14th Rock Mechanics Symposium, University of Pennsylvania, 85-106.

[18] Singh, D.P. 1975. A study of creep of rocks. Int. J. Rock Mech Sci & Geomech, 12, 271-276.

[19] Peng, S.S. 1973. Relaxation and the behaviour of failed rock. Int. J. Rock Mech Min. Sci. & Geomech., 10, 235-246.

[20] Kranz, R.L., and Scholz, C.H. 1977. Critical dilatant volume of rocks at the onset of tertiary creep. J. Geophys. Res., 82 (4), 893-4,898.

[21] Schmidtke, H., and Lajtai, E.Z. 1985. The long-term strength of Lac du Bonnet granite. J. Rock Mech. Min. Sci. Geo., 22, 461-465.

[22] Lau, J., Gorski B., Conlon, B., and Anderson, T. 2000. Long-term loading tests on saturated granite and granodiorite, Report: 06819-REP-01300-I 0016 ROO, CANMET, Natural Resources Canada.

[23] Cristescu, N.,D. 2009. Time effects in rock mechanics. In Proceedings of the SME Annual Conference, Albuquerque, New Mexico.

[24] Paraskevopoulou, C., Diederichs, M.S. 2013. Long-term behaviour in tunnelling: limitations in using creep parameters. In: Proceedings of 66th Canadian Geotechnical Conference, GeoMontreal on Geoscience for Sustainability, September 29–October 3, 2013, Montreal, Canada.

[25] Paraskevopoulou, C., 2018. Predicting the long-term behaviour of rock masses in tunnelling. In: Proceedings of the WTC 2018, Word Tunnelling Congress on The Role of Underground Space in Building Future Sustainable Cities, April 2018, Dubai, United Emirates.

[26] Paraskevopoulou, C., Perras, M., Diederichs, M.S., Loew, S., Lam, T., Jensen., M. 2018. Time-dependent behaviour of brittle rocks based on static load laboratory testing. Geotechnical & Geological Engineering, 36, 337-376(2018). https://doi.org/10.1007/s10706-017-0331-8

[27] Bukharov, G.N., Chanda, M.W., and Bukharov, N.G. 1995. The three processes of brittle crystalline rock creep. Int. J. Rock Mech. Min Sci. & Geomech, 32 (4), 325-335.

[28] Pellet, F., Hajdu, A., Deleruyelle, F. and Bensus, F. 2005. A visco-plastic model including anisotropic damage for the time-dependent behaviour of rock. Int. J. for Num. Anal. Meth. Geomech., 29, 941-970.

[29] Paraskevopoulou, C., Diederichs, M., 2018. Analysis of time-dependent deformation in tunnels using the Convergence-Confinement Method. Tunnelling and Underground Space Technology Journal, 71(2018), 62-80, 19 pages. http://dx.doi.org/10.1016/j.tust.2017.07.001

[30] Innocente, J., Diederichs, M.S., Paraskevopoulou, C, Aubertin, J.D. 2020. Numerical investigation of the applicability of time-dependent creep models. Submitted to the: Proceedings of the ISRM International Symposium – Eurock 2020 – Hard Rock Engineering, June 2020, Trodheim, Norway.

[31] Innocente, J., 2021. Time-dependency and long-term strength of rocks -limitations, interpretation and application in

finite difference continuum methods. MSc Thesis. In: Queen's University Publications. Canada.

[32] Paraskevopoulou, C., Perras, M., Diederichs, M.S., Amaan, F., Loew, S., Lam, T., 2015. Long-term static load laboratory testing behaviour of different rock types . In: Proceedings of the 68th Canadian Geotechnical Symposium GeoQuebec on Challenges from North to South. September 2015, Quebec, Canada.

[33] Paraskevopoulou, C., Perras, M., Diederichs, M.S., Amann, F., Loew, S., Lam, T., Jensen., M. 2017. The three stages of stress-relaxation - Observations for the long-term behaviour of rocks based on laboratory testing. Journal of Engineering Geology, 216, 56-75(2017). http://dx.doi.org/10.1016/j.enggeo.2016.11.010

[34] ASTM. 2013. Standard Test Methods for Stress Relaxation for Materials and Structures. E328-13. American Society for Testing and Materials.

[35] Paraskevopoulou, C., Perras, M., Diederichs, M.S., Amaan, F., Loew, S., Lam, T., 2015. Observations for the long-term behaviour of carboniferous limestone rocks based on laboratory testing. In: Proceedings of the EUROCK 2015 Future Development of Rock Mechanics – 64th Geomechanics Colloqium, October 2015, Salzburg, Austria.

[36] Paraskevopoulou, C., Perras, M.A., 2017. Investigating the long-term behaviour of brittle rocks: Visco-elastic creep parameters and time-to-failure. In: PRF 2017: Progressive Failure and Long-term Strength Degradation of Brittle Rocks ISRM Conference, June 2017, Ascona, Switzerland.

[37] ISRM. 2014. Suggested methods for determining the creep characteristics of rock.47, 275-290

[38] Goodman, R.E. 1980. Introduction to Rock Mechanics, John Wiley and Sons, New York.

[39] Davis, G.H., Reynolds, S.J., and Kluth, C.F. 2012. Structural Geology of Rocks and Regions. 3rd Edition. Wiley. New York. (ISBN:978-1-118-21505-0), 864 pages.

[40] Fairhurst, C., and Cook, N.G.W. 1966. The phenomenon of rock splitting parallel to the direction of maximum compression in the neighborhood of a surface. In: Proc. of the First Congress on the International Society of Rock Mechanics, National Laboratory of Civil Engineering, Lisbon, Portugal, 1, 687-692.

[41] Diederichs, M.S., Day, J.J., Ghazvinian, E., Perras, M., Paraskevopoulou, C., Walton, G., Progressive brittle damage processes and failure in rock. Keynote Lecture in the Progressive Failure and Long-term Strength Degradation of Brittle Rocks In: PRF 2017: Progressive Failure and Long-term Strength Degradation of Brittle Rocks. ISRM Conference, June 2017, Ascona, Switzerland.

[42] ISRM. 1979. Suggested methods for determining the uniaxial compressive strength and deformability of rock materials. 16, 2, 138-140.

[43] Martin, C.D. 1997. Seventeenth Canadian Geotechnical Colloquium: The effect of cohesion loss and stress path on brittle rock strength. Can. Geotech. J., 34 (5), 698-725.

[44] Martin, C.D. 1993. The Strength of massive Lac du Bonnet granite around underground openings. Ph.D. Thesis. University of Manitoba, Canada.

[45] Martin, C.D., Kaiser, P.K., and McCreath, D.R. 1999. Hoek-Brown parameters for predicting the depth of brittle failure around tunnels. Can. Geotech. J., 36 (1), 136-151.

[46] Perras, M.A., and Diederichs, M.S. 2014. A review of the Tensile Strength of Rock: Concepts and Testing. J. Geotech. and Geol. Eng., 32, 525-546.

[47] Gorski, B., Anderson, T., and Conlon, B. 2009. DGR site characterization documents, technical reports TR-08-11. Available at [www.nwmo.ca].

[48] Gorski, B., Anderson, T., and Conlon, B. 2010. DGR site characterization documents, technical reports TR-08-36. Available at [www.nwmo.ca].

[49] Diederichs, M.S. 2003. Rock fracture and collapse under low confinement conditions. Rock Mech. Rock Eng., 36 (5), 339-381.

[50] Panet, M. 1995. Calcul des Tunnels par la Me'thode de Convergence–Confinement. Presses de l'Ecole Nationale des Ponts et Chausse'es, Paris, 178 pages.

[51] Panet, M. 1979. Time-dependent deformations in underground works. In: Proc. of the 4th Int Congress on Rock Mechanics, Montreux.

[52] Sulem, J., Panet, M. and Guenot, A. 1987. Closure analysis in deep tunnels and Analytical solution for time-dependent displacement in a circular tunnel. Int. J. Rock Mech. & Min. Sci. & Geom. Abstracts. 24(3), 145-154 and 155-164.

[53] Paraskevopoulou, C., Diederichs, M.S., 2017. How time-dependency influences the Longitudinal Displacement Profile during the construction of deep tunnels. In: Proceedings of the WTC 2017, Word Tunnelling Congress on Surface Challenges – Underground Solutions, June 2017, Bergen, Norway.

[54] Paraskevopoulou, C., Diederichs, M.S., Perras, M. 2017. Time-dependent rock masses and implications associated with tunnelling, In: Proceedings of EUROTUN 2017, IV International Conference on Computational Methods in Tunnelling and Subsurface Engineering, April 2017, Innsbruck, Austria.

Section 3

Geotectonic

Chapter 3

Neotectonics and Stressed State Patterns of the Sakhalin Island

Leonid Bogomolov, Lidia Sim and Pavel Kamenev

Abstract

The study of neotectonics, neotectonic and modern stress of the Sakhalin has been performed by the set of methods. The scheme of modern geodynamics of the island has been constructed by the data of neotectonic activation of the faults. Three types of zones with dissimilar geodynamic conditions have been distinguished: transtension, transpression, and strike-slip (simple shift). The results of Sakhalin modern stress reconstruction based on focal mechanisms of earthquakes allowed to characterize the distribution of the stress state parameters over the island surface: the Lode-Nadai factor and the direction of axes of deviatoric compression and tension. The changes in characteristics of modern tectonic stress field have been noticed at the boundaries of regions with different regimes of modern faulting. Specific orientations of compression axes of the neotectonic stress field are proper for North Sakhalin. Therein, the directions of compression axes become northeast in contrast to the predominant sublatitude orientation on the island as a whole. The obtained data on neotectonics and inherited modern stress field are applicable to the problems of engineering geological support of oil and gas projects' realization in the Sakhalin (new wells construction, control of the pipelines stability, accompanying urban planning, etc.).

Keywords: Amur and Okhotsk microplates, the Sakhalin, fault zone, neotectonic and modern stress, slickensides, focal mechanisms of earthquakes, geodynamic regime, transpression, strike-slip

1. Introduction

Sakhalin Island belongs to the tectonically active region of Northeast Asia. Within its boundaries, the border between the largest tectonic plates of the Earth—the Eurasian Plate, North American and Pacific ones—passes through the island territory. A wide boundary zone represented by a set of independently moving microplates expands along the convergent boundaries of these plates.

The border between the Amur and Okhotsk (Okhotsk Sea) Plates, the largest microplates, is often associated with the Central Sakhalin Fault of meridional strike. Inside the Sakhalin, the interplate border is drawn along the Western Sakhalin Fault in the south, and as the arc line, partially coinciding with the Eastern Sakhalin Fault (**Figure 1**) [1, 2], in the Central and Northern Sakhalin. Alternatively, this border goes along the Central Sakhalin Fault (Tym-Poronai Fault in the Northern Sakhalin) [3].

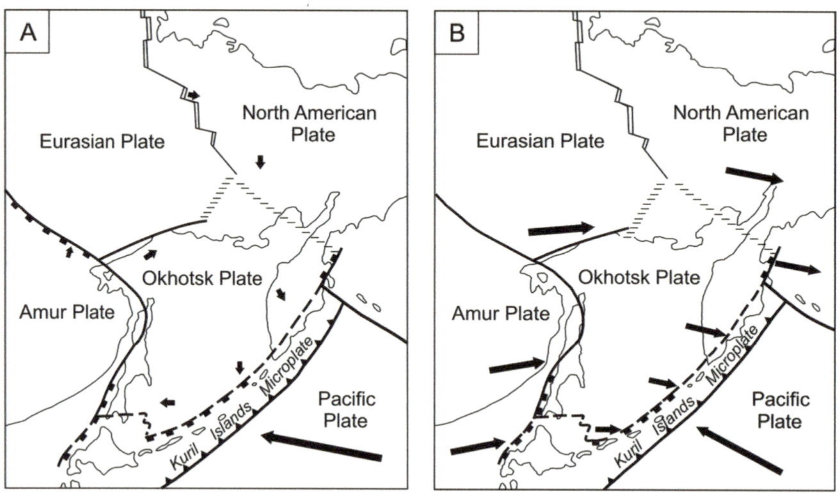

Figure 1.
Kinematics of the modern plate movements in the Okhotsk Sea region. (A) Relative to Eurasian Plate (not moving); (B) relative to Kuril Islands microplate. The arrows show the directions and values.

According to the paper [4], the interplate border is drawn along both the Western Sakhalin Fault and the Central Sakhalin Fault. The analysis of neotectonic stress to the West and East of the Central Sakhalin Fault allowed us to obtain new information about location of the border aforementioned [5–7].

In addition to our previous publications [5, 7], this chapter presents the results of extended tectonophysical studies based on a set of methods [6, 8–10], and the field measurements of 2019–2020 campaigns are involved. The paper presents the manifold manifestations of the geodynamic and seismic processes in the crust between the Amur and Okhotsk microplates, lying within the convergence zones of the Pacific, Eurasian, and North American tectonic plates.

The research aim is to study the recent tectonics, to reconstruct neotectonic stress by a set of methods, and to review the published seismological and geophysical data to approve or disprove the interplate border location. This involves the demonstration of effective but underused structural and tectonophysical methods to study the tectonics of some regions and to develop a model of the stressed state of Sakhalin Island. The above methods are able to give the characteristics of the post-Miocene stress field, but we will show (taking into account the seismological and GPS motion data) that the modern stress field in the Sakhalin crust is inherited mostly from the post-Miocene model. The modern stress nonuniformity as well as the climate change is proved to control both engineering geological processes and geotechnical conditions on the territory under consideration.

2. Structural-geomorphological method and the results

To reach the stated goal, the authors have compiled a structural-geomorphological (SG) map (neotectonics scheme) of Sakhalin Island using the method developed [11, 12], as well as a set of tectonophysical and structural methods, to study the modern stress state and kinematic types of individual faults. Thus, the patterns of neotectonic stress in the Northern and Central Sakhalin have been reconstructed by the structural-geomorphological (SG) method of shear stress reconstruction [13, 14]. The stress reconstruction has been performed also in the Southern Sakhalin by the following methods: the method of analysis of conjugate pairs of shear joints [8], the method of kinematic analysis of fracture structures [9], and the method of belts in the study of fracturing associated with discontinuous displacements [10].

Restored orientations of the axes of local stress states (LSSs) allow to build the unified single regional field of the Southern Sakhalin using the method [15]. The structural-geomorphological map of the Sakhalin is constructed on the grounds of the topography analysis using a topographic Sakhalin map of 1:500,000 scale. The longitudinal zonation of the newest structures is clearly shown on the map; the western and eastern zones of the latest uplifts are separated by the extended Central Sakhalin Depression. The Central Sakhalin Fault appears in the newest structure and serves as a border between the large Western ridge zone and the Central Sakhalin Depression. The zones of longitudinal strike are divided with the faults of various directions into a series of block-block uplifts, which differ in height. Some structures coincide with those that are marked in the work [16]. **Figure 2A** shows a selection of the newest structures.

The tectonic stress fields of Northern and Central Sakhalin have been reconstructed using medium-scale topographical maps and satellite images by the structural and geomorphological methods. This method envisages a special interpretation of megafractures—the small rectilinear relief elements, for which high density is an indicator of the lineament fault. The results of analysis of data on the orientations of subsidiary cracks in the shear zones have been summarized in [8] on the base of field simulation and shear zones mapping. If the mutual relative orientations of megafractures (two systems of joints and the ruptures, oriented along the bisector between them) correspond with the orientation of the subsidiary cracks in the shear zone (at this, they have a certain orientation to the fault plane), then the fault nature of a lineament is proved. Further, the orientations of compression and extension axes in the horizontal plane, the shift sign (right-left), and geodynamic conditions of the fault formation (transtension or transpression) are determined. The lower age limit of the fault activity is determined by the age of young Pliocene-Quaternary deposits, which are developed nearly throughout the study area and broken up with the megafractures and the latest faults. Neotectonic shear stress determinations by the faults, marked on the structural-geomorphological map, lend support to the validity of disjunction based on the terrane analysis. Different heights in the opposite walls of the fault allow to estimate the vertical movement component of displacement.

It should be noted that one of the horizontal axes can be an intermediate axis of the principal normal stresses.

The shear kinematic types of the faults correlate with the definitions of the kinematics of discontinuous faults mapped on the Northern Sakhalin [17, 18] as well as with seismic dislocations formed during the Neftegorsk earthquake (EQ) of 1995, Mw = 7.0 [19]. The reconstructed patterns of tectonic stress point to mainly submeridional extension and sublatitudinal compression over the considerable part of the island; but the compression axis orientation changes to northeast when the study zone moves northward (**Figure 2B** and **C**). The compression axes that turn to the northeast on the Northern Sakhalin is consistent with the scheme of the ellipsoid of the pre-late Miocene deformations of the entire Sakhalin presented in the work [18]. According to this work, the ellipsoid of deformations had turned around during the Pliocene-Quaternary time, the C axis or the shortening axis became latitudinal, which brought to a change in the kinematic type of shifts along the longitudinal meridional faults of the Sakhalin to the overthrust reverse fault type. Since the detailed research of the folded and discontinuous structures of different age performed by Rozhdestvenskii and Rozhdestvensky [17, 18] relate mainly to the Northern and Central Sakhalin, one can assume the change of stress state type cannot be applied to the Southern Sakhalin. The more so, the author notes

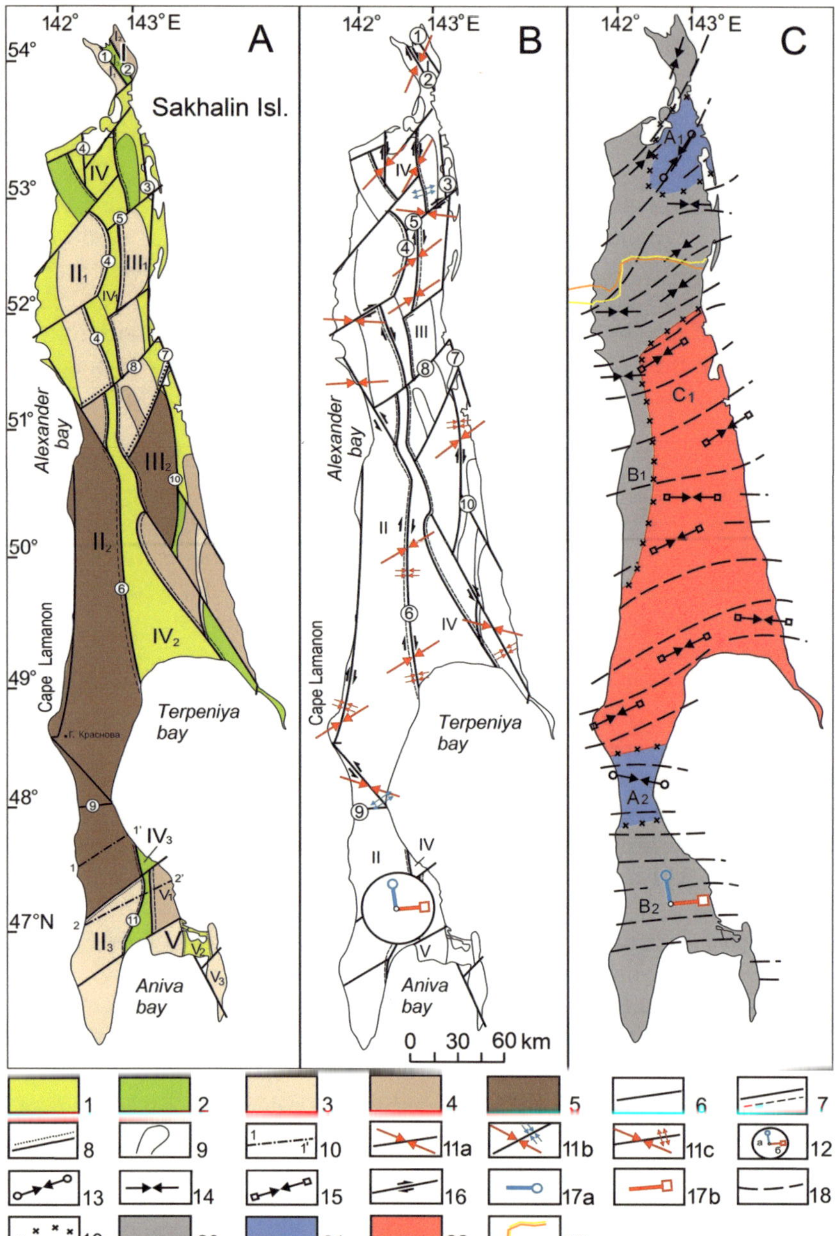

Figure 2.
Neotectonics and neotectonic stress of Sakhalin Island. (A) The map of Sakhalin Island, showing the structural and geomorphological features. 1–5: total syn-erosional uplifts; 1: less than 100 m, 2: 100–200 m, 3: 200–500 m, 4: 500–1000 m, 5: more than 1000 m, 6: the faults identified from the geomorphological data, 7: boundaries of the first-order structures, 8: boundaries of the second-order structures, and 9: the boundaries of uplifts and depressions. The digitals designate the uplifts on the map: I: Shmidt, II: West Sakhalin, III: East Sakhalin, V: Susunai, and IV: Central Sakhalin depression. Fault numbers are given in circles: 1: Central Sakhalin, 2: Hokkaido-Sakhalin, 3: Upper Piltun, 4: Nabilsky, 5: East Sakhalin, and 10: the geomorphological section line. (B) The scheme of neotectonic stress of Sakhalin Island. 11: Compression axes orientations in the horizontal plane and geodynamic conditions of their formation (reconstructed by the structural-geomorphological method); 11a: strike-slip, 11b: transtension, 11c: transpression, and 12: orientations of the subhorizontal extension (a) and compression (b) axes of the general stress field of the South Sakhalin. (C) The scheme of the latest geodynamics of Sakhalin Island. 13–15: Compression axes and geodynamic conditions; 13: transtension; 14: strike-slip; 15: transpression; 16: shift direction; 17: extension (a) and compression (b) axes reconstructed by the complex of field methods on the southern Sakhalin; 18: compression axes projections to a horizontal plane; 19: the boundaries of zones with different geodynamic regimes; 20–22: geodynamic regimes of 20: strike-slip, 21: horizontal extension (transtension), and 22: horizontal compression (transpression); and 23: oil and gas pipelines routes. The map of Bryantseva [5].

"appropriateness of the stress ellipsoid has been repeatedly questioned...," but its use has applied meaning, "although the anomalous structures, which are difficult to explain from this theory point of view, are observed" [18].

The zonation of the areas with different geodynamic conditions of faults generation within the latest stage has been conducted on the grounds of neotectonic stress, reconstructed for the Northern and Central Sakhalin (**Figure 2C**). General stress field of the Southern Sakhalin is shown below. Since the structural-geomorphological method is based on the patterns of mutual orientation of the subsidiary faults in the shear zone (by the data of simulation), the conditions of additional extension or compression, obtained in each certain case, need special additional studies to explain the cause of its occurrence. When shifts are simulating, the subsidiary fractures' different orientation to the actual direction of the shift has been obtained when additional compression or extension acted normally to the shear plane. Conducted zonation does not allow to separate the Western Amur and Eastern Okhotsk microplates (**Figure 2B**), as the compression axes' orientations on the Southern Sakhalin and south part of the Northern Sakhalin are uniform.

Fifty-six local stress states (LSSs) have been determined with the set of field methods on the Southern Sakhalin. These LSSs indicate the significant variance of the axes of principal normal stresses of local level (**Figures 3** and **4**), demonstrating the results of tectonophysical and structural methods' application. The method of jointing belt distribution [10] is shown in **Figure 3A** and **B**. The outcrop is located in the footwall of the Central Sakhalin Fault, whose strike is taken from the structural-geomorphological map (**Figure 2A**).

If the fault strike is known, but there are no data for its plane orientation in the space, then the method allows to determine this plane. To do this, the P1-P2 jointing belt with the pole at the P point is identified in the stereogram of fracture density, which is measured near the fault. The plane of the R1PR2 fault with the R pole is reconstructed by the connection of the fault strike points on the external circle of the R1R2 stereogram with the point of the P fracture belt pole. The S1 point (an intersection of the fault plane with the P1P2 belt) is a point of the displacement line outcrop across the fault plane on the upper hemisphere. We get the required displacement line across the fault by its connection with the stereogram center. Here, this line clearly tells about the strike-slip with some component of the thrust constituent. Since the fault is a dextral reverse one with a strike-slip component [18, 20], as well as by the relief pronounce, where the western wall is hanging, the displacement line has a sign of the dextral strike-slip fault with a reverse component. The strike-slip component dominance is obvious. The compression (σ_3^{gen}) and extension (σ_1^{gen}) axes are charted on the stereogram of the general stress field of the Southern Sakhalin (see below). According to the kinematic method [9], we can obtain the whole interval of possible displacements over the fault in this general field: from the S1 point by uniaxial extension to the S2 point by uniaxial compression, connecting the pole of the R fault to the σ_3^{gen} and σ_1^{gen} with the arcs of large circles. Thus, the displacement line, reconstructed by the analysis of the jointing distribution belt method, coincides with the direction of the strike-slip fault with a reverse component (actually the dextral strike-slip), if the stress field has been characterized by the compression condition. Two conclusions follow from above: (1) most of the fractures formed by belt distribution are meant to be tensile cracks that occur under extension conditions and (2) dextral strike-slip displacement has occurred along the fault at this stage. Two maxima of fractures density, outlined with the 8% isoline—I and II, are highlighted on the fractures belt (**Figure 3B**). The belt axis matches the intermediate axis σ_2; the direction of the compression axis, σ, and extension one, σ_1, are defined from the bisectors of the angles between I-I and II-II fractures according to the method of Gzovsky [8]. The directions of intermediate axis, σ_2,

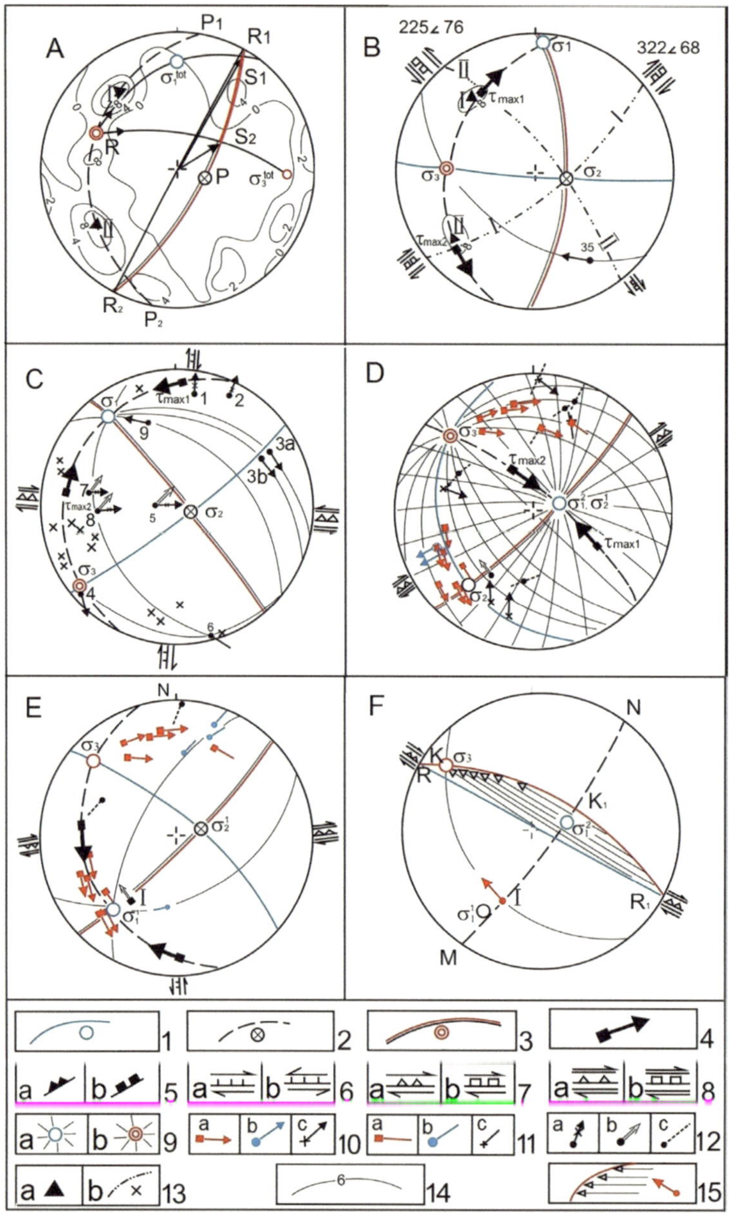

Figure 3.

Stereograms of the local stress state in the South Sakhalin. 1–3: Local stress state (LSS) axes and their action planes: minimal (σ_1); 2: intermediate (σ_2), maximum (σ_3) compression axes of the principal normal stresses; 4: poles of the planes of the maximum tangential stresses (τ_{max}) and displacement vectors; 5–8: strikes of the τ_{max} planes and fault kinematic types—5a: reverse faults, 5b: normal faults, 6a: dextral strike-slip faults, 6b: sinistral strike-slip faults, 7a: reverse faults with strike-slip component, 7b: normal faults with the reverse component, and 8: strike-slip faults with the reverse component; 9: strike-slip faults with the normal component; 9: arcs of the large circles divergent from—a: the extension axis and b: the compression axis; 10–13: displacement lines and vectors on the slickensides, which occurred due to—10a: uniaxial compression, 10b: uniaxial extension, and 10c: triaxial stress state; 11: the same, but with undefined displacement sign; 12: displacement lines—a: detected partial (vertical or horizontal) component of displacement, b: predicted, and c: contradict to the reconstructed LSS; 13: fractures—a: pole of the maximums of shear fractures and b: pole and plane of separate conjugated shear fractures or systems; 14: fractures density isolines; 15: pole and plane of the fault and displacement lines of the hanging wall. (A) and (B) Volume 16, the left bank of the Lyutoga river—(A) density and fracturing belt; (B) axes of the principal normal stresses; (C) volume 27, the right bank of the Vesely stream; (D–F)—South Sokolovskii quarry—(D) recent field, (E) ancient field of LSS in the hanging wall of the fault, and (F) the fault plane and displacement vector. Wulff-Gushenko modified grid, upper hemisphere.

Figure 4.
A slickenside on the outcrop of the Krasnoyarkovskaya suite K_2kr.

determined by two ways (both as the axis of the fractures belt and as the projection of the displacement line of two systems of the conjuncted shear fractures) coincide almost perfectly. This suggests the promising integration of different methods when studying the tectonic stress fields. According to the sum of displacements, the fault is a dextral reverse fault with a strike-slip component [20].

The unique point with the slickenlines in the Pleistocene alluvial-proluvial loams is located close to the Central Sakhalin Fault. Despite the small number of the slickensides, the local stress state (**Figure 3C**) of strike-slip faulting type with meridional extension and sublatitudinal compression axes has been reconstructed there. Measurement no. 1 and 2 are probably the traces of gravitational displacement along the slope.

Figure 3D and **E** show an example of the detection of two LSSs from the slickensides, measured in South Sokolovskii quarry, that has outcropped a melange of the West Susunai subterrane [21]. Most of the 37 displacement vectors have allowed to determine the LSS (**Figure 3D**), which is assumed to be more recent, as the most of slickenlines are the traces of latest displacements. The displacements vectors (slickenlines), which contradict this LSS, are marked with a special symbol, and their low integrity allows the LSS to be defined as more ancient.

The presence of clearly defined slickensides near the Central Sakhalin Fault points to high tectonic activity of the Southern Sakhalin (**Figure 4**). The outcrops of the Krasnoyarkovskaya suite K_2kr are presented with interbedding of tuffstones, sandstones, and tuff siltstones in the point of 47°01′51.00″ N and 142°30′00.40″ E coordinates. The slickensides originated with sinistral reverse fault with strike-slip component are shown in the right bottom part of **Figure 4**; the direction of motion of the lying wall of the fault is marked with an arrow.

Figure 5A and **B** show an example of detection of two LSSs of different ages as well as application of the method of fractures belt distribution by Danilovich to analyze the same planes with the slickensides.

Figure 6 presents the stereograms of the local stress state of the south part of the Central and South Sakhalin (56 determinations) charted on the scheme of the geological structure (according to Golozubov et al. [3], but simplified).

The unified regional stress field of this part of the island has been reconstructed previously from 56 determinations of the axes of the principal normal stresses in the South Sakhalin by the method justified in the works [14, 15]. The general field has the following angular characteristics: the extension axis ∠1–350∠10, the intermediate axis ∠2–112∠66, and the compression axis ∠3–260∠20. High-angle sinistral strike-slip faults with the dip azimuth of 32∠83 and dextral strike-slip faults with the dip azimuth of 125∠68 (**Figure 5E**) are the most frequent in this stress field.

A recent field studies has allowed to update the database on the local stress states both in the South Sakhalin and in the south part of the Central Sakhalin. All determinations are charted on **Figure 6**. The stereograms of different stressed state types

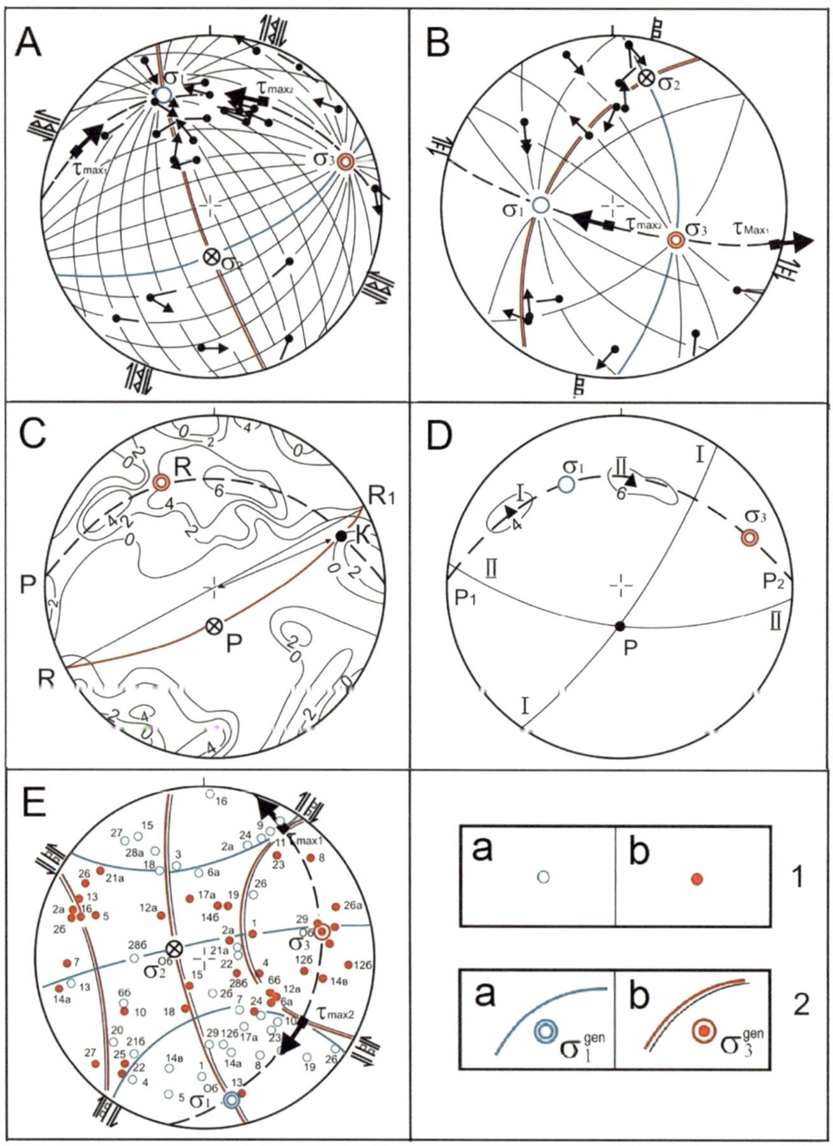

Figure 5.
Stereograms of tectonic stress and general stress field in the South Sakhalin. The axes of principal normal stresses of the LSS are designated with the symbols inside the frame 1: a—extension and b—compression. Designations of the general stress field axes and their planes are showed inside the frame 2: a—extension and b—compression. See the other designations in ***Figure 3****. (A)–(D) Limestone quarry and (E) the general stress field of the South Sakhalin.*

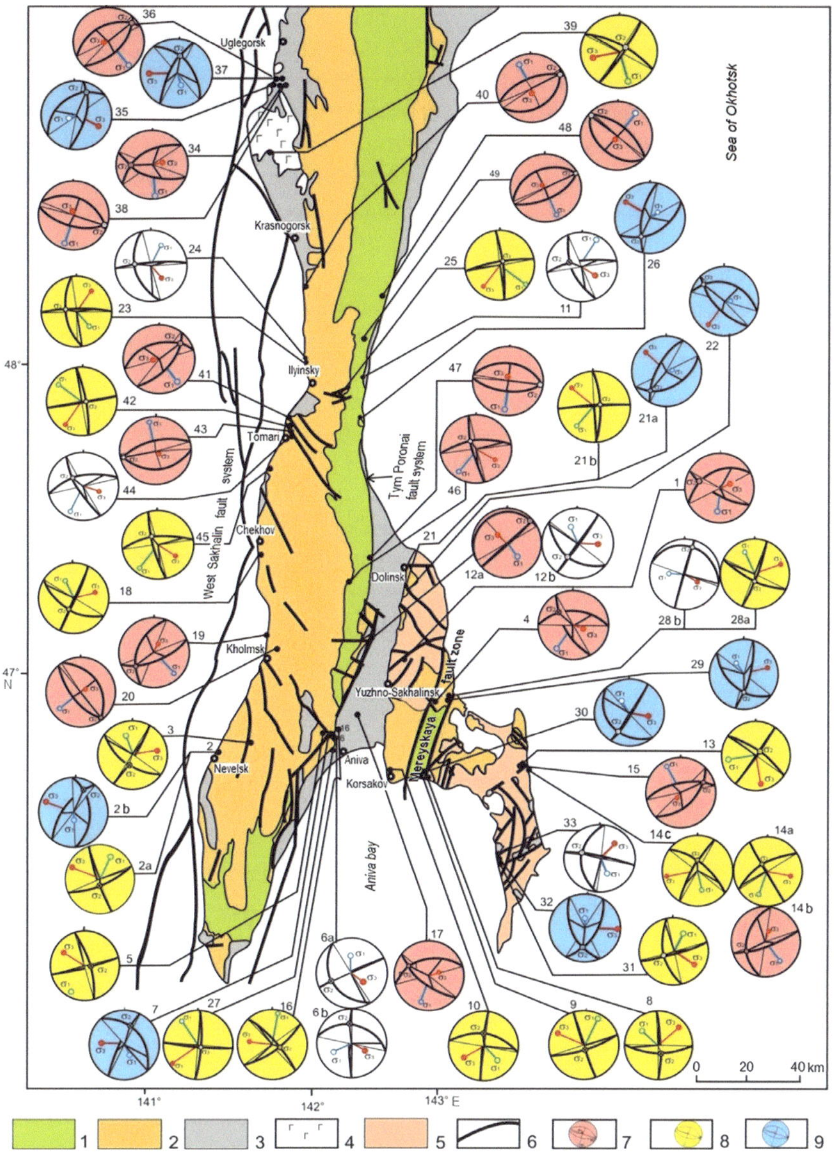

Figure 6.

The scheme of the geological structure (from [3], simplified) and local tectonic stress of the Southern Sakhalin (56 determinations), according to Sim et al. [5]. 1–5: heterochronous rock complexes; 1: Albian and upper cretaceous terrigenous, partly tuff-terrigenous, 2: Paleocene-Miocene terrigenous, volcanic are not general, 3: Miocene-quaternary terrigenous, 4: Pliocene basalts, and 5: Cretaceous and Paleocene-Eocene accretionary rock complexes of the Susunai and Tonino-Aniva terranes; 6: faults; 7-9: types of stereograms; 7: LSS of normal fault, possibly with strike-slip component, 8: LSS of reverse or oblique fault, 9: strike-slip fault LSS. The stereograms show the axes of principal normal (σ_1: minimum, σ_2: intermediate, and σ_3: maximum compressive) stress and the planes of maximum tangential stress with displacement vectors of the lacking wall (Wulff grid, upper hemisphere).

are highlighted with color. The stereograms of the LSSs of normal fault and normal fault with strike-slip component types are highlighted with reddish color and mean compression axis is to be oriented at an angle of 50–90° with the horizon; the LSSs of reverse fault and reverse fault with strike-slip component types are highlighted with blue color, and the extension axis is oriented at an angle of 50–90° with the horizon; the LSSs of strike-slip fault type, in which the compression and extension axes are oriented at an angle of 0–30°, are highlighted with yellow color; and the stereograms of the LSSs of unknown type, when all three axes are tilted 40–60°, are highlighted with white color. It is apparent that normal fault and normal fault with strike-slip component LSSs prevail over reverse fault with strike-slip component and reverse fault ones, which contradict the hypothesis of the post-fold change of the stressed state of strike-slip fault type to the reverse fault in the Sakhalin [7, 17].

Such hypothesis of work [7] was based on 15 LSS determinations only in the South Sakhalin. The later studies involving 56 points of LSS determinations as well as the analysis of seismicity (see below) have proved the prevailing strike-slip stress field both in fold and orogenic stages.

Explicit indicators of submeridional extension and sublatitudinal compression have been revealed on newly occurred marine terrace in the southern part of the town of Nevelsk. This terrace is a peculiar case of outcrops considered, because it resulted from the coseismic seabed uplift during Nevelsk earthquake (EQ) of August 2, 2007, M = 6.2 [22]. The vertical displacement of the earth surface reached 1.2 m near the coastline. The new marine terrace (actually the drained bench, **Figure 7A**) is located on the west wall of the West Sakhalin Fault, at 15–20 km distance from the Nevelsk EQ hypocenter. Two systems of cleft joints are well expressed in this bench in the Lower Miocene laminated shales with interlayers of silica marls. The post-Miocene field of strike-slip fault type, with horizontal extension and compression, has been reconstructed on the basis of the rose diagram of

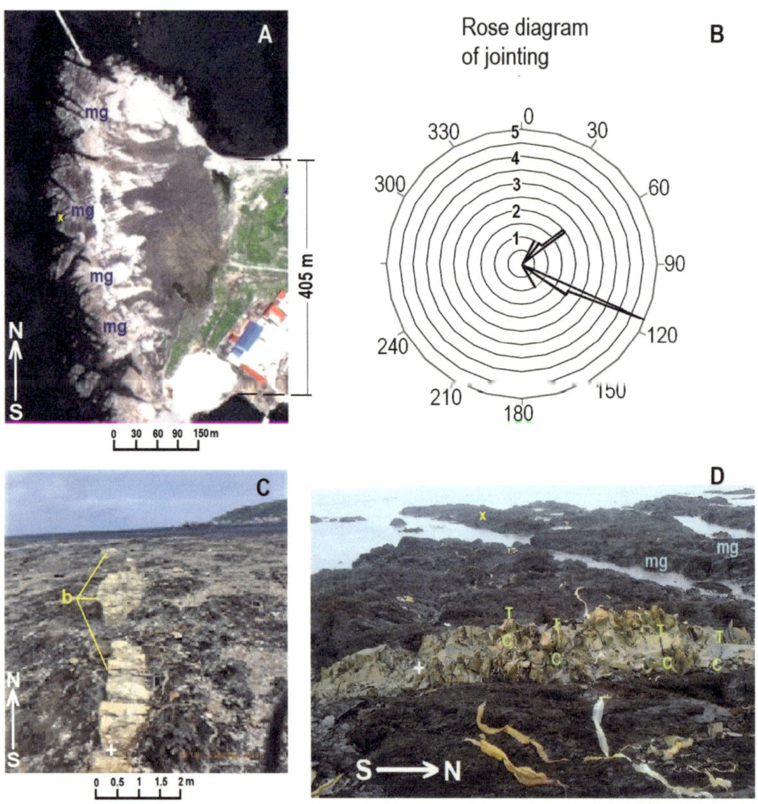

Figure 7.
*Indicators of post-early Miocene stresses on the newly formed marine terrace (dried bench in the southern part of the town of Nevel'sk). (A) Image of the terrace with a scale after wwwyandex.ru/maps (2014) (dark is sea surface with visible minigulfs, mg); (B) rose diagram of jointing on its surface (the compression axis by bisector of acute angle after Gzovsky [8] is oriented along azimuth of 92.5°, and the extension axis is oriented along the azimuth of 182.5°); (C) view of three boudins on the dried bench; and (D) relative position of a boudin and two minigulfs, mg, and tension cracks (indicated by T–C characters) in the boudin b4 specified in **Table 1**. Yellow "x" symbol denotes a point of reference near coastline, white "+" is a point of reference inside b4 boudin.*

# boudin	Length, m	Width, m	Distance to the next boudin, m	Extension, %	Remark
b1	3.8	0.55	3	69	First line, 30 m East from the water's edge
b2	4.8	0.95	4.6	80	
b3	1.4	0.65	8.3	405	
b4	10.3	1	13.2	117	
b5	2.8	1.05	8.2	213	
b6	4.3	1.25	29.2	526	
b7	7.1	1.3	41	488	
b8	5.1	0.95	23.5	388	
b9	4.1	0.5	2.16	47	
				259	Mean over first line
b10	2.16	0.65	2.2	78	Second line, 24 m to east of the first
b11	1.4	0.35	0.7	40	
b12	1.3	0.35	7.4	448	
				267	Mean over second line
b13	7.4	1.7	5.9	65	Third line, 23 m to east of the second
b14	Fragmented and partially under water				
b15	1.3	0.92	1.5	68	Fourth line,
b16	>1.6	Partially under water			26 m from the third
				217	Overall mean

Table 1.
The parameters of the boudins on the newly occurred marine terrace and the estimates of tensile strain.

jointing for these data (**Figure 7B**). The method of Gzovsky [8] has been used for this reconstruction.

The boudins in the layer of silica marls (**Figure 7C**) as well as the tension cracks in the boudins (**Figure 7D**) have confirmed surely that the extension axis is horizontal and its direction is close to the north-south one. We have measured the length and width of each boudin and the distance between neighboring ones in the direction of elongation to evaluate the maximal strain. The results are represented in **Table 1**. The level of tensile strain has been evaluated as the ratio of this distance to the half sum of their lengths. After averaging, we come to the estimate of post-Miocene extension of nearly 200%.

3. Discussion: correspondence with geophysical data

Generally, our obtained data on the absence of significant difference in stressed state in the West and East Sakhalin do not contradict the results of the global project "The World Stress Map" applied to the region of the North West Pacific, that is, the Sakhalin surroundings [23]. This project takes into account the data on Sakhalin earthquakes' focal mechanisms that are relevant to the pattern of the recent tectonic stress and strain. Although the statistics of such focal mechanisms was scanty in [23], the obtained map with the border between Amur and Okhotsk microplates

(**Figure 8**) is of special interest. One can see in **Figure 8** that this border is close to the location of the Central Sakhalin Fault (Tym-Poronai Fault) in the Central part of the island. Its location in the Northern Sakhalin corresponds well with that of Upper Piltun and Nabilsky Faults, see **Figure 2A**.

The southern part of the microplates boundary in **Figure 6** diverges with the West Sakhalin Fault. This aspect is not consistent with our approach, and we take into consideration additional regional data. The spatial distribution of the aftershocks of strong earthquakes, Gornozavodskoe EQ, 17.08.2006, M = 5.9, and Nevelsk EQ, M = 6.2, occurred nearly in the West Sakhalin Fault [22], which gave extra reason that this fault had labeled the debated boundary. It is more vital that the detailed analysis of the earthquake focal mechanisms, carried out in [23], has demonstrated rather a lateral zonation of the orientation of modern tectonic stress.

A convincing argument for the dominant regime of horizontal compression and strike-slip is the results [24] obtained from the data on the deep borehole drilling in the north and south of the Sakhalin. It has been shown in this work that the maximum sublatitudinal compression may exceed the vertical stress by 1.2–4 times (on average) both in the north and south of the Sakhalin. The caliper logging data have demonstrated the horizontal stress (sublatitudinal compression) to predominate over the vertical one. There are significant wellbore breakouts of the studied vertical holes in two antipodal angular sectors pointing the direction of the maximum stress-strain effect in a number of studied vertical boreholes in the Northern and Southern Sakhalin.

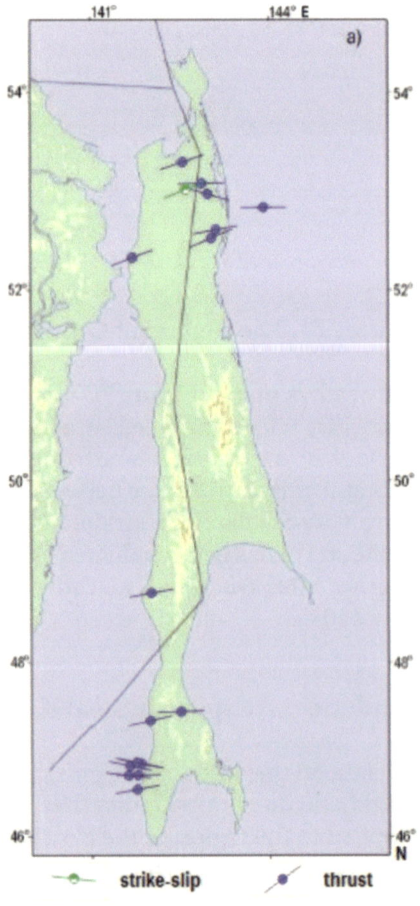

Figure 8.
The World Stress Map data regarding to the Sakhalin, according to Heidbach et al. [23].

Besides, the data treating of IFZ-19 profile of deep seismic sounding [15], which transected (crossed) South Sakhalin from the southwest to the northeast, gave tilted reflecting area M1 lying in 40–60 km depth and descending toward Okhotsk Sea. Meanwhile, there is no large fault in the Sakhalin Island, which could be associated with the border between the Amur and Okhotsk microplates. It is correspondent with the assumption from the work [20], that the Central Sakhalin Fault is an eastern branch of the Western Sakhalin Fault with the westward dip and lesser propagation depth, while the main Western Sakhalin Fault has the eastward dip. The simulations of interaction of tectonic plates surrounding the Sakhalin Island, which are based on GPS displacement surveys [25], have shown that one can draw the most likely border between the Amur and Okhotsk microplates along the Western and the Central Sakhalin Faults on the Southern Sakhalin territory, and along the Upper Piltun and Nabilsky Faults (**Figure 2A**) on the Northern Sakhalin. Seismological data, namely, the significant concentration of earthquakes hypocenters on the western coast of Sakhalin Island and in the Tatar Strait [25], can also testify that the border between the microplates passes along the Western Sakhalin Fault, while the Central Sakhalin Fault is its branch ending at the minor depth.

The analysis of tilt of the P and T axes by the catalog of focal mechanisms of the earthquakes occurred in 1962–2011, published in [26], has shown the obvious non-uniformity of orientations of these axes, see **Figure 9**. The maxima of the tilt angles of both P and T axes are 60–70°. This implies the reverse and normal faults. The strike-slips have a subordinate meaning. The graphs in **Figure 9** are bimodal, with two maxima of the P and T axes tilt distribution, and this prevents the assumption that the recent stress field is characterized as a reverse faulting type.

So, the actual results mentioned above contradict the conclusions about the border between the Amur and Okhotsk microplates passing along the Central Sakhalin Fault, which have been made on the grounds of the field two-dimensional tectonophysical studies using the ellipsoid of deformations. Those studies have demonstrated the different tectonic forces orientation in the Sakhalin, induced by differently directed movement of the tectonic plates, specified in [3]. On the Northern Sakhalin, the explicit maximum of earthquakes epicenters concentrates in the eastern part of Shmidt Peninsula and may correspond to the border between the microplates (**Figure 1**). Besides, the distribution graphs of the P and T axes tilt angles (**Figure 9**) also contradict the conclusion about the change in the regime of Sakhalin Island tectonic evolution within the folding stage (i.e., the regime of strike-slip type modified to the reverse faulting during the following orogenic stage).

The reconstruction of recent stress by the data of earthquakes focal mechanisms was carried out by Savvichev and published in the works [5, 6]. There are the orientations of maximum deviatoric extensions on the left, and compressions ones are on the right (**Figure 10**). Obviously, the western part of the island is more seismically active in comparison with the eastern one due to the border between the Amur and

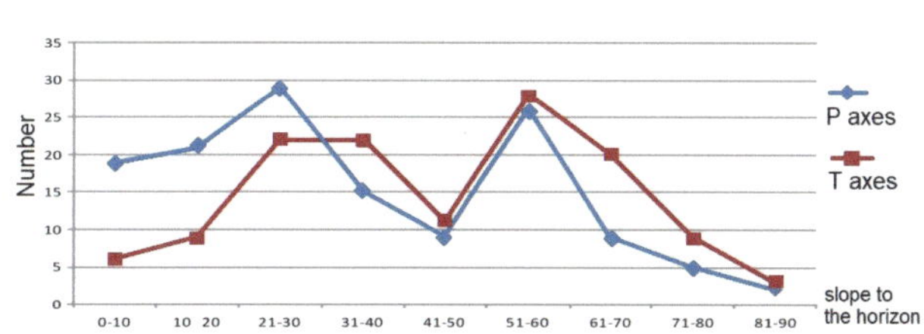

Figure 9.
The graphs of P and T axes tilt angles according to the earthquakes' focal mechanisms (1962–2011) by the catalog [26]. 135 determinations of the P and T axes in total, according to Konovalov et al. [26].

Okhotsk microplates, which passes along the Western Sakhalin Fault. Seismicity of the Northern and west of the Southern Sakhalin is considerably more active than in the central part of the island.

In this connection, the NE orientations of the subhorizontal axes of maximum deviatoric compression dominate in the north of the Northern Sakhalin, and they change the orientation to the ENE southward, up to 53°N. The first latitudinal band of nonstable orientation of the stress axes is distinguished directly southward of 53°N. The section with vertical orientation of the compression axis engages the attention in this band, indicating that the local geodynamic regime is a horizontal compression with latitudinal direction. The focal mechanism of destructive Neftegorsk earthquake in 1995 [19, 27] is able to confirm such regime. The second band of nonstable orientations of the stress axes is distinguished between 51° and 52°N in the northern part of the Central Sakhalin. The isolated group of the earthquake focal mechanisms with a subhorizontal latitudinal orientation of the maximum deviatoric compression and steeply descending south-eastward extension ones is distinguished in the western part of the Central Sakhalin between 49.5–50.0°N and 142.0–42.9°E. This approximates to the geodynamic regime of horizontal compression. The western part of the Southern Sakhalin is characterized with stable orientations of horizontal compression, which become less stable, changing the orientation to the WNW and ENE, as the Poyasok Isthmus is approached. In the same northern part of the Southern Sakhalin, the steep orientations of the maximum deviatoric extension in the west become unstable when approaching eastward, where the steep dip angles are noted in the central part and the slow ones to the SE in the eastern part. Variability of the orientations of recent deviatoric extensions and compressions is evidently associated with the boundaries of different geodynamic zones distinguished during neotectonic stress consideration.

Resuming, we note a reasonable correspondence of the presented results on orientation of the principal compression and extension axes with the results of studies of the earth surface strains in the vicinity of active faults of Sakhalin Island on the base of GPS/GLONASS surveys data [28]. In accordance with this work, the

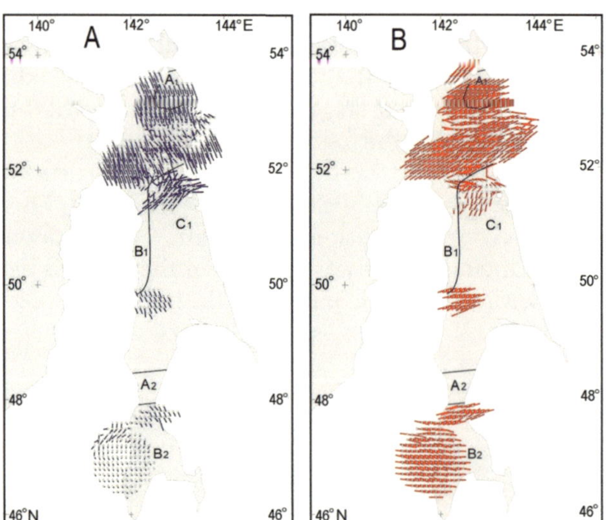

Figure 10.
The projection of principal maximum deviatoric extensions (A) and compressions (B) to the horizontal plane. The point of dip vector origin is marked with a circle; when the dip angles are less than 15°, the circle locates in the middle of the dip vector. The areas of different geodynamic conditions are marked with black lines (see ***Figure 2C****), according to Sim et al. [5].*

GPS/Glonass surveys for three cross sections (on the northern, central, and southern parts of Sakhalin Island) provided initial information on the horizontal velocities of the GPS displacements and the error of their determination, which allowed to estimate strain of the earth's surface of Sakhalin Island. The GRID_STRAIN software suite [29] has been used for computations. The GPS-observation points location, vectors of the average annual horizontal velocities, and the computed horizontal deformations of surface in the northern, central, and southern parts of Sakhalin Island are shown in **Figure 11**. As can be seen in **Figure 11B**, the shortening of the earth's crust of Sakhalin occurs mainly in the sublatitudinal direction, and it slightly varies from region to region. The heterogeneity of the surface deformation field is appeared in the distribution of the principal elongation and shortening axes over the Sakhalin area (the terms elongation and shortening are used to characterize the strain field and extension and compression for the stress field).

The territory of the northern part of the island (**Figure 11B**, upper frame) is subjected to compression in a southwestern direction. Maximum velocities of deformation up to 13×10^{-8} per year appear in its eastern part. Directions of the deformation velocities, in general, are consistent with the orientation of the compression and extension axes in the restored field of tectonic stress of the Northern Sakhalin

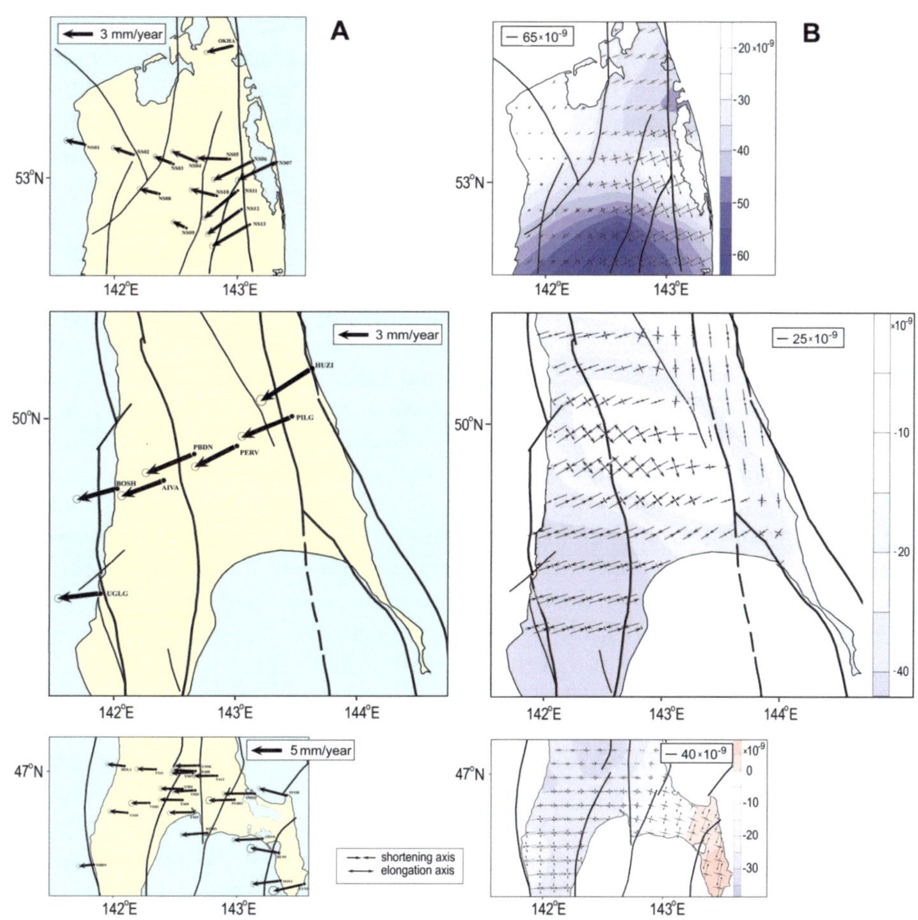

Figure 11.
Horizontal velocities at the GPS observation points of the Sakhalin Island in relation to the Eurasian tectonic plate (A), and velocities of dilatation and main axes of earth surface deformation (B). Upper frame: the measurements in the northern part of the Sakhalin during 2003–2013; midframe: the same in the central part in 2000–2011; and bottom frame: the same in the southern part in 1999–2009, according to Prytkov and Vasilenko [28].

(**Figure 2C**) [5, 6]. However, according to the GPS measurements, no extension area was found in this part of the island as opposed to **Figure 2C** with an extension zone to northward of 53°N. Low values of the velocities of deformation of the earth surface, not exceeding ~5×10^{-9} per year (**Figure 11B**, midframe), are typical for the central part of the island. Hokkaido-Sakhalin Fault delimitates the territory with different geodynamic conditions: the southwestern direction of the compression axes turns to the submeridional orientation at the fault's boundary. Alongside the dominant sublatitudinal shortening, prevailing over most of the Southern Sakhalin territory, the extension area is distinguished to eastward of 143°E (**Figure 11B**, the bottom frame). The maximum velocity of deformation of shortening is ~8×10^{-9} per year. In the vicinity of the Central Sakhalin Fault, the axes of shortening change their western direction to northwestern, and the compression of the north-eastern direction becomes a prevailing regime of deformation. This zone has not been revealed when reconstructing of neotectonics stresses due to the insufficiency of initial data for the LSS determination.

Therefore, at the subregional scale (100 km and more), one can see the correspondence of the results obtained using three methods: structural-geomorphological, geophysical (seismological data), and GPS measurements. Discrepancies are observed for the zones of ~30 km length, they may be related to the complexity of fault structure on the Sakhalin, including multiple local breaks [16–18, 21], as well as to very short period of seismic and GPS monitoring. Partially motivated by this correspondence, we foresee the results represented in **Figure 2C, 6, 10** and **11** to be used for several regional issues of engineering geology works and developments. The information about the direction of maximal horizontal stress action is extremely important for new wells' construction in the oil and gas fields in the Northern Sakhalin (including the Sea of Okhotsk shelf). This action may cause wellbore breakouts followed by negative incidents: casing collapse or stuck pipe. The safe directions of near horizontal wells drilling are determined considering stress conditions. In the case of dangerous azimuth of the well, they (the operators) can compensate the horizontal stress bias by drilling mud treatment like the mud weight increase, the use of oil-based mud and inhibition additives. The data we obtained may be used for this technology in addition to the limited information of the WSM project data [23], and this will contribute to the technological safety of wells drilling and production.

Characteristics of the modern stress field controlling the earth surface deformations are significant also for the geological support of railway and main roads operation. These communications run in the vicinity of Central Sakhalin Fault. Moreover, the fault zone in the center of South Sakhalin is the most populated area in the island. The boundary between the microplates implies a zone of enhanced seismicity (see the paragraph after **Figure 8**), and seismic hazards for civil engineering. Therefore, the particular result that the segment of this boundary in the Southern Sakhalin does not follow the Central Sakhalin Fault is crucial for detailed seismic zoning and seismic microzoning before urban development. Finally, we remark that the pipelines from Sakhalin oil and gas terminals to the continent cross the active faults of submeridional strike. Our result about the location of microplates' boundary in the Northern Sakhalin speaks in favor of that the pipelines stability control to be focused on the section crossing eastward faults (Hokkaido-Sakhalin Fault, Upper Piltun one, etc.) rather than the West and Central Sakhalin Faults.

4. Conclusion

Zonation of the subregions of various geodynamic conditions of the latest faults formation has been originally carried out on the Sakhalin on the base of tectonophysical and structural-geomorphological (SG) methods. The boundaries of the mapped zones have evinced a variability of the modern stress field parameters. Reconstruction of the recent tectonic stresses of the Sakhalin has revealed a dominance of strike-slip type of stress state with subhorizontal compression and extension axes. The compression axis is sublatitudinally oriented, and the extension is submeridional. Sublatitudinal compression axes, reconstructed for the Central and Northern Sakhalin by the two-dimensional SG method, are turning to the northeast. The computed parameters of tectonic stresses as a whole are consistent with the results obtained from the earthquake focal mechanisms and GPS/GLONASS movements of earth's surface. The results of this work have demonstrated that the border between the Amur and Okhotsk microplates along the Central Sakhalin Fault on the Southern Sakhalin and change of strike-slip stress field during the folding stage to the reverse fault one to be inappropriate. The further research by SG and other tectonophysical methods of stress field reconstruction may result in the specified map of geodynamical regimes and geohazards in the Sakhalin. Such updated map will not be so complicated as structural geological one in **Figure 2A**. However, the visible indicators of local stress state are rare on the studied territory due to the forest cover. Therefore, the next research stage is expected to involve wider processing of satellite images to get raw data on faulting for the neotectonic and modern stress field reconstruction.

Acknowledgments

The authors thank G. Bryantseva, A. Gorbunov, V. Degtyarev, and V. Parovyshny for the successful participation in the field works and/or the assistance and O. Zherdeva for preparing graphical materials.

Author details

Leonid Bogomolov[1*], Lidia Sim[2] and Pavel Kamenev[1]

1 Institute of Marine Geology and Geophysics, FEB RAS, Yuzhno-Sakhalinsk, Russia

2 Schmidt Institute of Earth Physics, RAS, Moscow, Russia

*Address all correspondence to: bleom@mail.ru

References

[1] Savostin L, Verzhbitskaya A, Baranov B. Modern tectonics of the Okhotsk Sea region. Doklady of the Academy of Sciences of the USSR. Earth Science Sections. 1982;**266**(4):62-67

[2] Seno T, Sakurai T, Stein S. Can the Okhotsk plate be discriminated from the north American plate? Journal of Geophysical Research: Solid Earth. 1996;**101**(B5):11305-11315. DOI: 10.1029/96JB00532

[3] Golozubov V, Kasatkin S, Grannik V, Nechayuk A. Deformation of the upper cretaceous and cenozoic complexes of the West Sakhalin terrane. Geotectonics. 2012;**46**:333-351. DOI: 10.1134/S0016852112050020

[4] Vasilenko N, Prytkov A. GPS-based modeling of the interaction between the lithospheric plates in Sakhalin. Russian Journal of Pacific Geology. 2012;**6**:35-41. DOI: 10.1134/S1819714012010137

[5] Sim L, Bogomolov L, Bryantseva G, Savvichev P. Neotectonics and tectonic stresses of the Sakhalin Island. Geodynamics & Tectonophysics. 2017;**8**:181-202. DOI: 10.5800/GT-2017-8-1-0237

[6] Sim L, Bryantseva G, Savvichev P, Kamenev P. Patterns of transition zone between Eurasian and North American plates (by example of stressed state of the Sakhalin island). Geosystems of Transition Zones. 2017;**1**:3-22. DOI: 10.30730/2541-8912.2017.1.1.003-022

[7] Sim LA, Bogomolov LM, Kuchai OA, Tataurova AA. Neotectonic and modern stresses of South Sakhalin. Russian Journal of Pacific Geology. 2017;**11**:223-235. DOI: 10.1134/S1819714017030058

[8] Gzovsky M. Fundamentals of Tectonophysics. Moscow: Nauka; 1975. p. 536. (in Russian)

[9] Gushchenko O. The method of kinematic analysis of destruction structures in reconstruction of tectonic stress fields. In: Grigoriev A, Osokina D, editors. Fields of Stress and Strain in the Lithosphere. Vol. 1979. Moscow: Nauka. pp. 7-25. (in Russian)

[10] Danilovich V. The Method of Belts in Studies of Fracturing Associated with Shearing. Irkutsk: Irkutsk Polytechnic Institute; 1961. p. 47. (in Russian)

[11] Kostenko N. Development of Folded and Faulted Deformation in Orogenic Relief. Moscow: Nedra; 1972. p. 320. (in Russian)

[12] Kostenko N, Briantseva G. On issue of structural-geomorphological interpretation in closed spaces environments. Bulletin of Moscow University. Series 4: Geology. 2004;**4**:34-38. (in Russian)

[13] Rebetsky Y, Sim L, Marinin A. From the Slickensides towards Tectonic Stresses. Methods and Algorithms. GEOS: Moscow; 2017. p. 234. (in Russian)

[14] Polets A. The stress state of the Sakhalin Island and adjacent territories. IOP Conference Series: Earth and Environmental Science. 2019;**324**:012010. DOI: 10.1088/1755-1315/324/1/012010

[15] Filonenko V, Pavlenkova G. Some features of the dynamics of the crust in the Okhotsk Sea region. In: Solodilov L, editor. Geophysics of the XXI Century: 2005: Proceedings of the V.V. Fedynsky Seventh Geophysical Readings. Moscow: Nauchny Mir; 2006. pp. 101-107. (in Russian)

[16] Voeikova O, Nesmeyanov S, Serebryakova L. Neotectonics and Active Faults of Sakhalin. Moscow: Nauka; 2007. p. 187. (in Russian)

[17] Rozhdestvenskii V. Role of shifts in formation of the Sakhalin structure. Geotectonics. 1982;**4**:99-111. (in Russian)

[18] Rozhdestvensky V. Active rifting in the Japan and Okhotsk seas and the tectonic evolution of the Central Sakhalin fault in the Ceinozoic. Russian Journal of Pacific Geology. 2008;**2**:15-24. DOI: 10.1134/S1819714008010028

[19] Rogozhin E, Reisner G, Besstrashnov V, Strom A, Borisenko L. Seismotectonic settings of Sakhalin Island. Izvestiya Physics of the Solid Earth. 2002;**38**:207-214

[20] Trifonov V, Kozhurin A. Study of active faults: Theoretical and applied implications. Geotectonics. 2010;**44**:510-528. DOI: 10.1134/S0016852110060051

[21] Zharov A. Geological Structure and Cretaceous-Paleogene Geodynamics of South-East Sakhalin. Yuzhno-Sakhalinsk: Sakhalin Publishing House; 2004. p. 191. (in Russian)

[22] Tikhonov I, Kim C. Confirmed prediction of the 2 August 2007 MW 6.2 Nevelsk earthquake (Sakhalin Island, Russia). Tectonophysics. 2010;**485**:85-93. DOI: 10.1134/S1028334X08040417

[23] Heidbach O, Rajabi M, Cui X, Fuchs K, Müller K, Reinecker B, et al. The World Stress Map database release 2016: Crustal stress pattern across scales. Tectonophysics. 2018;**744**:484-498. DOI: 10.5880/WSM.2016.001

[24] Kamenev P, Bogomolov L, Zakupiv A. On the stress state of the Sakhalin crust according to the data of drilling deep boreholes. Russian Journal of Pacific Geology. 2017;**11**:25-33. DOI: 10.1134/S1819714017010043

[25] Bulgakov R, Ivaschenko A, Kim C, Sergeev K, Strel'tsov M, Kozhurin A, et al. Active faults in northeastern Sakhalin. Geotectonics. 2002;**36**:227-246

[26] Konovalov A, Nagornykh T, Safonov D. Modern Studies of Earthquake Focal Mechanisms in the Sakhalin. Vladivostok: Dal'nauka; 2011. p. 252. (in Russian)

[27] Rogozhin E. Focal mechanism of the Neftegorsk, Sakhalin earthquake of May 27 (28), 1995. Geotectonics. 1996;**30**:124-131

[28] Prytkov A, Vasilenko N. Earth surface deformation of the Sakhalin Island from GPS data. Geodynamics & Tectonophysics. 2018;**9**:503-514. DOI: 10.5800/GT-2018-9-2-0358

[29] Teza G, Pesci A, Galgaro A. Grid_strain and grid_strain3: Software packages for strain field computation in 2D and 3D environments. Computers and Geosciences. 2008;**34**:1142-1153. DOI: 10.1016/j.cageo.2007.07.006

Chapter 4

Multivariate Assessment of California Bearing Ratio with Contrasted Geotechnical Properties of Soils in Ilorin-Lokoja Highway

Attah Fakeye, Olusegun Ige and Olufemi Ogunsanwo

Abstract

California Bearing Ratio (CBR) is an important parameter used in designing pavement layers in road construction but testing this parameter requires time, labor, and huge cost. The study therefore applies multivariate approach to evaluate CBR based on contrasted geotechnical parameters along Ilorin-Lokoja highway. The results obtained showed that the migmatite-gneiss-derived soils are slightly more fines (< 0.075 mm; 7.4–59.6%), more plastic (PI; 1.6–39%), and have low strength (MDD = 1.8 mg/m^3; CBR = 29.0%) than the metasediments (11–57.7%, 2.0–30%, 1.6 mg/m^3, 23.6%) and older granite soils (8.2–32.7%, 2.6–13.4%, 1.7 mg/m^3, 27.8%), respectively. The principal component analysis (PCA) revealed three major components (eigenvalues >1) which accounted for 83.8% of the total variance at the rate of 33.4, 14.7, and 11.4%. Major contributing variables for the components were fines (R = 0.87), plasticity index (R = 0.7), and coarse sand (R = 0.67%). Spatial distribution of these groups established interplay of sediment-gradation and moisture-connection evident in hierarchical cluster analysis that revealed patterns of homogeneity and soil relationships. Regression analysis established five models from predictor variables such as fines, activity, free swell, liquid and plastic limits, weighted plasticity index, optimum moisture content, and maximum dry density with the coefficient of determination (R^2 = 0.33) and root mean square error (RMSE) of 7.80.

Keywords: multivariate, principal component analysis, regression, hierarchical analysis, geotechnical properties

1. Introduction

Identification and quantitative characterization of soils are of dire importance in geotechnical assessment despite the difficulties experienced using conventional approach. Index properties are important parameters in the analysis of geotechnical engineering problems, particularly to estimate strength of the soil material. Conversely, laboratory test takes 2–4 days to measure compaction and California Bearing Ratio (CBR) values for pavement design. As a result, they are expensive and time-consuming.

Also due to lack of specialized personnel, these tests are oftentimes avoided in many soil investigation programs. Thus, the need to incorporate statistical approach in predicting soil properties becomes inevitable.

Several authors have applied this approach in relating and predicting soil properties. One to one relationship was presented among soil properties [1] such as liquid limit (LL), plastic limit (PL), plasticity index (PI), optimum moisture content (OMC), and maximum dry density (MDD). Furthermore, Carter and Bentley explained that soil type, density, moisture content play an important role in soil relationship [2] and correlated soil expansion index and plasticity index, fine fraction and weighted plasticity index (i.e., product of PI and percentage passing 0.425 mm). Apart from index properties, some researchers like Owoseni et al. [3] and Yildrin and Gunaydin [4] observed that California Bearing Ratio depends on other factors such as type of soils, permeability of soil, maximum dry density and optimum moisture content. To correct overlapping problem and uncertainty in prediction, Yitagesu et al. applied multiple regressions to improve the ability of predicting soil properties, and better model the extent of their relationship [5].

This paper attempts to identify geotechnical characteristics of soils developed on different rocks and establish relationships among various properties in order to estimate soil strength capability in three lithological units. Multivariate approach using principal component analysis (PCA) and hierarchical classification methods are used to identify patterns, detect and classify new parameters into groups; and further propose regression models to determine CBR values in view of huge cost and labor.

2. Methodology

Soil samples (130 samples) were collected along the Ilorin-Lokoja highway (>300 km length) which spans across latitude 7°25′N-8°40′N and longitude 4°30′E-6°45′E. Simultaneously, Garmin GPS was used to record coordinates of sample locations. The topography ranges from a relatively flat to hilly, undulating terrain with elevation ranging between 100 and 700 m above sea level. The highway is overlying the Precambrian Basement rock of South Western Nigeria (**Figure 1**) and cut across three geologic units: the migmatite-gneiss complex (denoted by PCB), the metasediments/volcanic series (PCM) and the older granite series (PCG) [6].

Majority of the rock is the migmatite—gneiss essentially made up of migmatite and banded gneiss. Others are flaggy quartzite with biotite gneiss, undifferentiated schist, porphyritic granite (porphyroblastic), and medium-coarse grained biotite and hornblende granite. Temperature ranges from 25 to 35°C. Climate is dry to wet, with a mean annual rainfall of 1200 mm. Due to heavy rainfall, considerable moisture change occur in the soils which dries up at prolonged dry season. This induces soils susceptibility to volume changes.

2.1 Laboratory analysis

Geotechnical tests were carried out on air-dried (35–40°C) soil samples at the Soil Geotechnical Laboratory of Nigerian Building and Road Research Institute following the British Standard [7] Part 2: Clause 9.2, 4.5, 5.3, 5.4, and Part 4, Clause 3.3 and 3.4 methods. The soil engineering parameters obtained include natural moisture content, Atterberg limits, particle size distribution, free swell, compaction test and California Bearing Ratio.

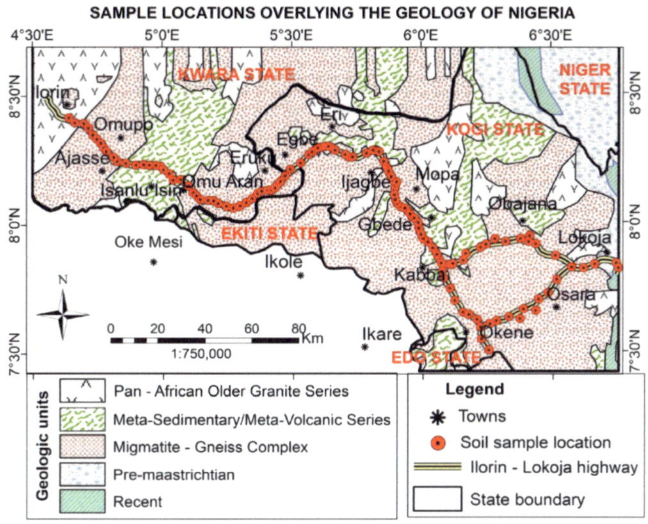

Figure 1.
Geology of Nigeria showing the study highway overlain by sampling points (red dots).

2.2 Statistical analysis

In SPSS statistical software, 20 soil parameters were explored and their relationships examined. Data transformation was applied to ensure equal influence on the model thus, fulfilling the linear model assumptions. The strength and relationship trends on the dataset were examined from Pearson correlation matrix and quantitative measures of linear associations determined. Principal component analysis (PCA) approach was incorporated to reduce the data with many variables, identify clusters, and transform the soil variables into new uncorrelated variables that preserve most of the information [8]. Components with eigenvalues >1 were retained and subjected to varimax rotation to maximize correlation between the factors and measured variables. Thereafter, Agglomerative Hierarchical Cluster (AHC) analysis was computed to identify analogous behavior among different soil characteristics and soil individuals using Ward's method and squared Euclidean distance as a measure of similarity between soils [9].

3. Results and discussions

The statistical summary of the laboratory test is shown in **Table 1**. The soils exhibited wide variations of data clustering around the mean value (1.08–88.6%) and high coefficients of variation (1.7–147%). The median of some parameters was lower than the mean value, indicating a low effect of abnormality on sampling values.

3.1 Particle size characteristics

In the migmatite-gneiss derived soils (PCB), gravel and coarse sand varied with coefficient of variation (CV) from 18.3 to 100% (<23.5%), medium to fine sand was between 8.0 and 86% (>32%) while the percentage of silt and clay were 3.2–50.8% (52.4%) and 0.9–34.6% (58.8%) respectively. However, the percentage of fines (<0.075 mm) ranged between 7.4 and 59.6% (48.0%). This proportion of

Properties	Units	Migmatite-Gneiss (PC3) = 87					Metasediment/Metavolcanic (PCM) = 33					Older Granite (PCG) = 11				
		Min	Max	Mean	SD	CV	Min	Max	Mean	SD	CV	Min	Max	Mean	SD	CV
NMC	%	2.6	22.7	11.8	4.5	38	3.1	18.4	12.1	4.1	34	2.4	16.2	7.5	4.3	56.8
Gravel		38.6	100	89.3	13.2	14.8	65	100	87	10.5	12.1	45	100	85.8	19.6	22.8
CS		18.3	99.2	80.4	18.9	23.5	43.1	100	74.6	15.3	20.6	19.4	100	77.5	25.8	33.3
MS		13.6	86	51.5	16.6	32.3	21.5	84.2	53.8	15.3	28.5	18.9	92.1	53.9	18.8	34.9
FS		8	79.5	28.9	14.3	49.4	15.2	71.1	36.8	13.3	36.1	14	41	28	7.9	28
Sand		20	87	59.5	14.3	24	25	81	50.1	13.9	27.8	31	81	57.6	17.4	30.2
Silt		3.2	50.8	17.1	9	52.4	6.1	44.3	20.3	8.4	41.2	2.4	24.5	15.5	7.5	48.6
Clay		0.9	34.6	12.3	7.3	58.8	1.8	28.7	15.2	7	46.3	1.2	18.8	10.8	6.6	60.8
Fines		7.4	59.6	23	11	48	11.7	57.7	28.6	10.4	36.5	8.2	32.7	21.8	7.5	34.6
Ac		0.1	8.5	1.1	1.2	110	0.2	7.9	0.9	1.4	147	0.2	5.1	1.2	1.4	123
Fsw		1.8	28.4	6.4	5.7	88.3	2.8	19.6	7.1	5.1	72.2	3.2	4.7	4	0.5	11.6
LL		13.4	69	28.4	11.7	41.1	13.4	48.5	30.3	9.7	32	16.5	32.4	23.2	5.7	24.5
PL		2.2	50	19	9.6	50.5	10.2	31.7	20.3	6.3	31	8.6	26.5	16.6	5.7	34.5
PI		1.6	39	9.4	6.5	68.7	2.1	30.1	10	7.3	72.6	2.6	13.4	6.5	3.4	51.9
wPI		0.3	31.9	5.1	4.6	90.6	0.9	25.3	5.8	5.7	97.8	0.9	7.3	3.5	2.1	60.6
BD	mg/m^3	1	2.9	2	0.3	15	1.5	2.2	1.9	0.2	8.5	1.7	2.1	1.9	0.1	5.9
DD		0.8	2.4	1.7	0.2	13.9	1.2	1.9	1.5	0.2	10.8	1.5	1.9	1.6	0.1	8
MDD		0.9	2.6	1.8	0.3	14.4	1.3	2.1	1.6	0.2	11.9	1.6	1.9	1.7	0.1	7.6

		Migmatite-Gneiss (PCB) = 87					Metasediment/Metavolcanic (PCM) = 33					Older Granite (PCG) = 11				
MC	%	7.6	27	15.9	4.8	29.8	7.6	23.9	16.5	3.8	23	9.1	19.7	14.1	3.6	25.8
OMC		5.7	25	13	4.2	32.2	10.1	22.5	15	2.9	19.6	8.9	18.2	14	2.9	20.8
CBRu		12.5	70	50.8	12.4	24.4	17	75.1	47.7	11.9	25	30.4	58.6	52.5	8	15.3
CBRs		10	56.4	29	9.7	33.4	11.2	45	23.6	7.1	30	12.1	37.2	27.8	6.5	23.3
SP		1.5	15.4	5.5	2.8	51.9	2	11.6	6	3	49.8	2.3	7.9	4.3	1.6	38.4
Dr		0.7	1	0.9	0	4.6	0.9	1	0.9	0	3.2	0.9	1	1	0	1.7
Wr		0.8	2	1.3	0.2	19.7	0.8	1.6	1.1	0.2	20.9	0.6	1.6	1	0.3	26.8
LLr		0.7	6.3	2.7	1.1	42.8	1.3	9.1	2.9	1.7	58.3	2	7.8	3.9	1.9	48.3
PIr		1.1	8.1	1.7	1	56.2	1.1	2.6	1.5	0.4	25.7	1.1	2	1.5	0.3	21.5

Parameter abbreviations are described in the Abbreviations section

Table 1.
Statistical summary of soil properties.

fines is similar to those reported by Ige et al. [10]. In the metasediment derived soils (PCM), content of gravel and coarse sand were higher from 43 to 100% (<20.6%), while medium to fine sand was between 15.2 and 84.2% with 36% CV. Similarly, the percentage of silt and clay ranged between 6.1–44.3% (41.2%) and 1.8–28.7% (46.3%). The proportion of fines (11.7–57.7%; 36.6% CV) is relatively as high as the PCB origin. Similarly, the older granite rock (PCG) exhibited a wide range of gradation with gravel and coarse sand ranging between 19.4 and 100% (<33.3%). Medium to fine sand content was lower (14–92%) (CV = 28–34.9%) while percentage of silt and clay varied between 2.4–24.5% and 1.2–18.8% (CV = 48.6–60.8%), respectively. The amount of fines (8.2–32.7%) and CV (34.6%) are very low in this area.

On one hand, this granularity is similar to the work of Nwaiwu et al. [11] where the lateritic soils are enriched with gravel and sands ranging between 28.2–40% and 42.2–48% resp. However, the high percentage passing through No. 200 (0.075 mm) BS sieve suggests the soil is predominantly of fine materials and classified according to Unified Soils Classification System (USCS) system as clayey sand (SC), silty sand (SM) and silty, clayey sands (SC-SM). Other soil classes obtained include poorly graded sand with silt or clay (SP-SM, SP-SC), poorly graded gravel with clay or silty clayey gravel (GP-GC, GC-GM), silty gravel (GM), sandy lean or fat clay, (CL, CH), and sandy silt or elastic silt (ML, MH) that occurred in low percentage. Similarly, according to American Association of State Highways and Transportation Officials (AASHTO) system, the most dominating classes are A-2 and A-7 soils, hence rated as excellent to good and fair to poor materials for road use.

3.2 Consistency limits

A wide range of plasticity (**Figure 2**) characterized the inorganic silty clayey soils in the area. The liquid limit varied between 13.4 and69% with a lower range experienced in PCG derived soils (<32.4%), while plastic limit and plasticity

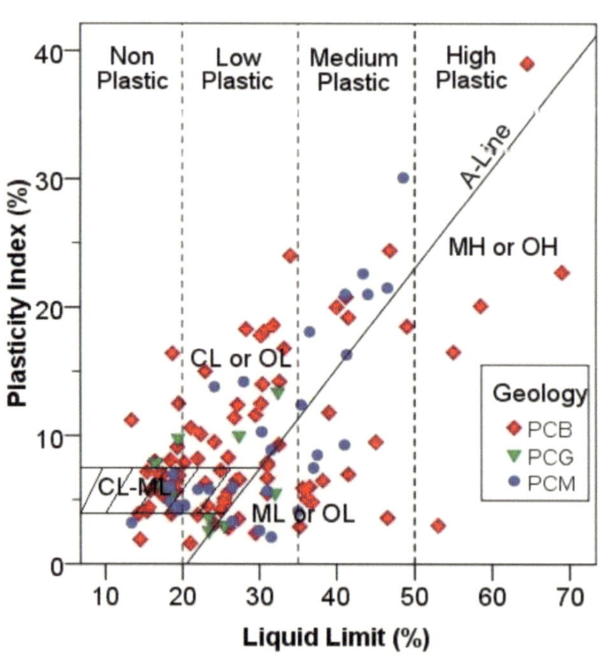

Figure 2.
Casagrande chart of plasticity-liquid limit relationship.

index at the PCB unit ranged between 2.2–50% and 1.62–39%, with mean values of 28.4, 19.0, and 9.4%, respectively (**Table 1**). The Casagrande plasticity chart revealed majority of the soils from the migmatite-gneiss origin placed above the A-line, indicating that they are composed of inorganic clay material and exhibited low to medium plasticity, implying low to medium swelling and compressibility. The moderate plasticity suggests low to medium dry strength, which could easily crumble under load thus leading to pavement failure and possible erosion under climatic threat. The distribution of the soil samples on the chart portrayed the variability in soil plasticity characteristics.

Moreover, free swell (Fsw) varied from 1.8 to 28.4% in PCB, 2.8–19.6% in PCM and 3.2–4.67% in PCG with mean values ranging from 6.4, 7.1, and 4.04%, respectively; while soil activity with mean values oscillated between 0.09–8.5 (1.1), 0.17–7.85 (0.9), and 0.2–5.08 (1.2) within the 3 units. The weighted plasticity index (wPI) value ranged between 0.25–31.9% (5.1%), 0.92–25.3% (5.8%), and 0.87–7.3% (3.5%) with mean from the 3 units. In PCB soils, activity tends to be higher than normal (8.5), high weighted plasticity index (wPI), plasticity ratio (PIr), and swelling potential (SP) indicating that the soils are active. The result of natural moisture content (NMC) (2.6–22.7%) is fairly high, considering the time of sample collection. This indicates the soil potential for water retention, which is a property of fine-grained soils. The high water content also suggests the presence of high water table earlier reported by Adams et al. [12]. These observations correspond with Bayamack et al. [13]. The derived plasticity parameters (wPI, PIr, SP, and LLr) represent the effective contribution of the plasticity of fines to the performance of the entire soil materials, depending on the amount of fines.

3.3 Compaction and California Bearing Ratio

The maximum dry density (MDD) of the soils from PCB area (**Table 1**) increases with mean to 2.6 mg/m^3 (1.77 mg/m^3) at 25% (13%) optimum moisture contents (OMC). These values are higher than those obtained in metasediment (PCM) and older granite (PCG) units with 2.1 mg/m^3 (1.6 mg/m^3) MDD and 22.5% (15%) OMC. The low density-moisture relationship implies low strength instigated by loose soils that are susceptible to erosion. The interaction of the subgrade with water greatly reduces strength and therefore promotes continuous failure of the overlying pavement. Few examples of soil compaction curves (**Figure 3**) illustrate distinct peak of maximum dry density at optimum moisture content.

The CBR values at 95% OMC after 48 hours of immersion varied between 10 and 56.4% for PCB, 11 and 45% for PCM and 12.1 and 37.2% for PCG soils (**Table 1**). The mean values within the three lithological units varied between 28.8, 23.6, and 27.8%, respectively. For unsoaked condition, the CBR varied in a higher rate from 12.5 to 75.0% within the 3 units. The result showed a reduction in strength due to soaking suggesting a probable drastic reduction in strength by more than half during wet condition and the penetration resistance becomes reduced due to excessive moisture. These values are similar to those found along Ado Ekiti-Akure road (27–100%) by Adams and Adetoro [14]. The low mean CBR value (<30%) suggests that the soils may not withstand ground vibrations when vehicular load is applied and reinforces its susceptibility to erosion. Soil improvement measures are therefore, envisaged for the stability of soils for adequate strength.

3.4 Simple linear regression

High statistically significant correlation (R > 0.70) is recorded among 13 soil attributes pairs (**Table 2**) such as gravel, coarse sand (CS), medium sand (MS), fine

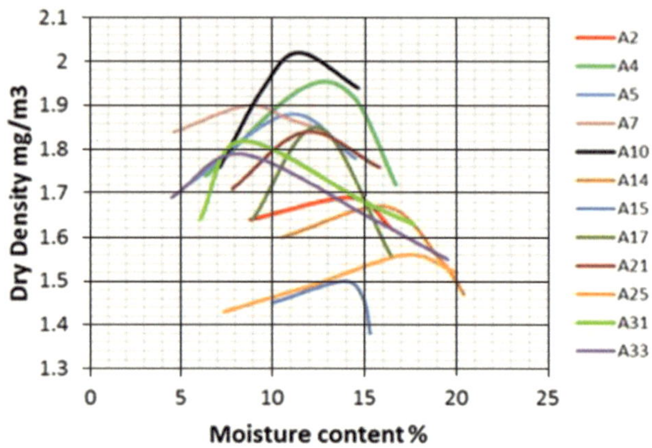

Figure 3.
Compaction curves of selected soil samples.

Properties	R values	Properties	R values
FS and fines	0.97	Fsw and CBRu	-0.62
Gravel and CS	0.96	Fsw and PI	0.61
BD and MDD	0.95	LL and PI	0.61
DD and MDD	0.95	NMC and clay	0.60
Silt and clay	0.94	MS and silt	0.60
Silt and fines	0.92	LL and MC	0.59
BD and DD	0.90	PL and MC	0.56
PI and wPI	0.89	PI and MC	0.56
Clay and fines	0.88	FS and sand	-0.56
Fsw and LL	0.88	NMC and LL	0.56
FS and silt	0.80	MS and clay	0.56
LL and PL	0.80	DD and OMC	-0.55
FS and clay	0.77	MDD and OMC	-0.54
CS and MS	0.75	Gravel and sand	0.52
MC and OMC	0.72	fines and OMC	0.52
Gravel and MS	0.70	FS and LL	0.52
LL and CBRu	-0.70	FS and PL	0.51
MS and FS	0.68	FS and OMC	0.51
MS and fines	0.68	NMC and Fsw	0.51
Fsw and wPI	0.68	Clay and Ac	-0.51
Fsw and PL	0.66	fines and LL	0.51
LL and wPI	0.64	NMC and fines	0.50
PL and CBRu	-0.62	Sand and fines	-0.50

Table 2.
Pearson significant correlation of soil properties.

sand (FS), silt, clay, fines, swelling potential (SP), free swell (Fsw), liquid limit (LL), plasticity index (PI), dry density (DD) and maximum dry density (MDD) which raises the issue of multi-collinearity. However, other parameters exhibit low correlations (R < 0.50) including sand, activity (Ac), plastic limit (PL), moisture content (MC) and optimum moisture content (OMC).

This could be attributed to the presence of high fine fractions and potential influence from environmental factors. The result corroborates with the observations obtained on gneiss derived laterite in Central Cameroun [15] and reaffirms the views of the earlier scholars that geotechnical properties of laterites depends on the parent materials, climate, vegetation, topography and duration of the laterization phenomenon [16].

3.5 Multivariate analysis of soil properties

3.5.1 Principal component analysis

Among the multivariate analysis techniques, principal component analysis is the most frequently used because it is the starting point in data mining which aims at minimizing the dimensionality of the data. Seven principal components (PCs) were extracted with eigenvalues >1 which accounted for 83.8% of the total variance of data (**Table 3**).

However, the first five PCs accounted for >70% of variability in measured soil properties. While PC1 explained 33.4% of the total variance with fines as the major contributing variable (R = 0.87), PC2 accounted for additional 14.7% of the total variance with plasticity index (PI) as the second major contributing variable (R = 0.70). In PC3, 11.4% was accounted for, with coarse sand (CS) contributing more (R = 0.67). Other components accounted for <15% and as such were removed as they explained less variance than individual variable in the dataset [8].

Based on the communality estimates, the five factors explained more than 90% of variance in MDD, PI, LL, DD, BD, FS, CS, SP, fines and gravel; > 80% in wPI, PL, Fsw, Wr, LLr, NMC, sand, silt and clay; > 70% in MS, MC, CBRu, and Dr.; above 60% in PIr, OMC, and Ac; and 53% in CBRs (**Table 4**). According to Johnson and Wichern [17], a high communality suggests that a high proportion of the variability is explained by the factor with a higher preference over a low communality estimate. By implication, the factors fairly explained the variance in soaked CBR and as such required a regression model to predict the property. The values obtained are similar to those obtained by Shukla et al. [18].

The coefficient of linear correlation between the variables and their factors (**Table 4**) give a meaning to the principal components. The parameters are well represented and explained by the factorial axes on the correlation circle (**Figure 4**).

		Initial eigenvalues	
PCs	**Eigen values**	**% of Variance**	**Cum % of variance**
1	9.0	33.4	33.4
2	3.9	14.7	48.1
3	3.1	11.4	59.5
4	2.5	9.2	68.7
5	1.6	5.8	74.5
6	1.4	5.2	79.6
7	1.1	4.2	83.8

Table 3.
Eigenvalues and proportions of variance explained by PCA.

Soil variables	PC1	PC2	PC3	PC4	PC5	Communalities
SP	0.93	—	—	—	—	0.91
PI	0.87	—	—	—	-0.39	0.95
LL	0.87	—	—	—	0.33	0.94
Fsw	0.86	—	—	—	—	0.85
wPI	0.84	—	—	0.34	—	0.89
CBRu	-0.60	—	—	—	-0.32	0.74
MDD	—	0.96	—	—	—	0.97
DD	—	0.94	—	—	—	0.94
BD	—	0.93	—	—	—	0.91
OMC	0.35	-0.60	—	—	—	0.68
MC	0.43	-0.56	0.32	—	—	0.76
LLr	—	—	-0.78	—	0.39	0.82
Ac	—	—	-0.75	—	—	0.68
C	—	—	0.72	0.39	—	0.85
NMC	0.48	—	0.73	—	—	0.80
S	—	—	0.67	0.42	—	0.86
Fines	0.32	—	0.61	0.43	0.36	0.96
FS	0.36	—	0.53	0.40	0.42	0.91
G	—	—	—	0.94	—	0.94
CS	—	—	—	0.94	—	0.95
MS	—	—	—	0.78	—	0.79
S	—	—	-0.41	0.47	-0.47	0.85
PIr	—	—	—	—	-0.77	0.68
PL	0.43	—	—	—	0.72	0.87
Wr	—	—	—	—	0.89	0.82
Dr					0.83	0.71
CBRs	0.39	0.37			0.48	0.53

Table 4.
Proportion of variance and communality estimates of soil variables.

This graph shows three groups of variables, suggesting the existence of correlation between them.

PC1 positively correlates (> 0.84) with SP, PI, LL, Fsw, wPI and MC, NMC, PL, CBRu (**Table 4**) and is termed plasticity parameters. PC2 demonstrated very high positive correlation with soil densities (MDD, DD, BD) (> 0.93) and negatively correlated with moisture contents (MC, OMC) (<-0.6) and is termed moisture-density or compaction parameters since the variables are important functions of soil moisture density. It also showed moderate positive loading from CBRs (0.37) resulting from significant correlation between MDD and OMC. Similarly, PC3 defined as fine gradation parameters showed highest positive correlation (0.72) with clay and NMC; FS, silt and fines (0.53, 0.61, 0.67); and negatively correlated (>0.75) with activity and liquid limit ratio (LLr). These variables are a function of fine soil texture. PC4 and PC5 are positively correlated (>0.70) with coarse materials (gravel, coarse and medium sand), referred to as coarse soil texture.

3.5.2 Agglomerative hierarchical analysis

Agglomerative hierarchical clustering (AHC) of the principal components showed a representation of the soil variables in homogenous classes where soils of same class exhibited similar values. The three major classes distinctively categorized by the AHC are depicted by dendrogram (**Figure 5**) which also displayed the cohesiveness of clusters formed. Class 1 soils were characterized by fines (silt, clay), plasticity and moisture contents (MC, OMC) parameters. Approximately, 47% of these soils were of the migmatite-gneiss derived origin (PCB) with a p-value <0.001.

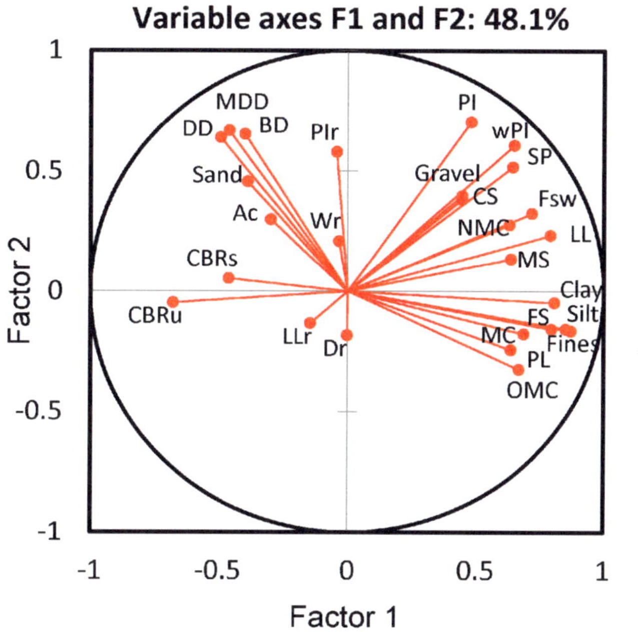

Figure 4.
Score plot and correlation circle obtained with PCA.

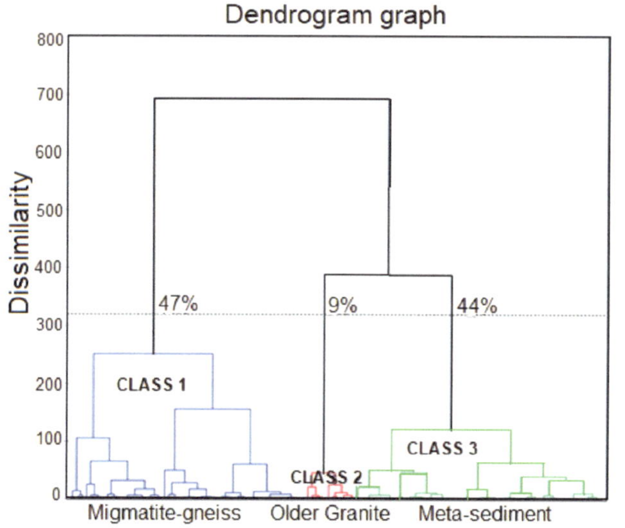

Figure 5.
Dendrogram of the studied soil variables.

Similarly, Class 3 showed significant clustering with densities (BD, DD, and MDD) parameters of mainly metasediment origin (PCM). Class 2 soils performed poorly owing to few parameters and p-value >0.05, which might be from the older granite derived origin (PCG).

3.5.3 Multiple regression analysis

Following stepwise regression method, five models were generated in (Eqs. 1–5). The result indicated that between 25 and 33% of the variation in soil properties was explained by the combination of these predictors. In Eq. 5, 70% training dataset accounted for 33% variance with coefficient of determination ($R^2 = 0.33$) and root mean square error of performance (RMSE = 7.8). Given the p-value <0.001 computed by analysis of variance (ANOVA), the significance level (5%) and the low bias (0.05), the prediction by the explanatory variables is significant.

$$CBRs = 0.008LL - 0.29PL - 0.5wPI - 0.1OMC + 6.87MDD + 25.2 \quad (1)$$

$$CBRs = 6.47MDD - 0.014LL - 0.32PL - 0.59wPI - 0.12OMC + 0.12Fsw + 26.7 \quad (2)$$

$$CBRs = 8.51MDD - 0.1LL - 0.2PL - 0.56wPI - 0.1OMC + 0.11Fsw + 38.1Dr - 0.13Wr - 0.2LLr + 0.88PIr - 14.02 \quad (3)$$

$$CBRs = 0.062LL - 0.51PL - 0.82wPI - 0.28Fsw - 38.06 \quad (4)$$

$$CBRs = 0.31Fines + 1.88Ac + 0.41Fsw - 0.298LL - 0.25PL - 0.73wPI - 0.5OMC + 2.11MDD + 36.03 \quad (5)$$

4. Conclusion

All the variables exhibited a large variation of data clustering around the mean value and high coefficients of variation. The soils within the area are predominantly very fine sands with a high percentage passing through No. 200 (0.075 mm). The natural moisture content and soil activity are moderately high which may be due to soil potential for retaining water. The result of Atterberg limits shows most of the soils classified as inorganic clayey soils under A-2 and A-7 groups. Their variability is locational and between lithology which reflects the influence of parent materials and pedogenic activities. Application of correlation analysis has allowed for the determination of the relationship between index properties, compaction and CBR and for deriving multivariate relationships for the assessment of CBR based on these parameters. Strong correlation existed among index properties but showed weak relationship with CBR under soaked condition. Principal component analysis categorized the measured soil parameters into five major groups for which first three components explained more than half of the total variance. Hierarchical classification validated the clustering of different individuals/variables based on the parent material. Five empirical models were obtained between soaked CBR and index parameters. The coefficient of determination (R^2) and root mean square error (RMSE) revealed that the models obtained were able to predict the target variable to a good degree of accuracy. The pavement challenges witnessed on the highway are attributable to the poor subgrade, the influence of geology and lack of drainage. It is therefore recommended to stabilize the soil with cement or lime in order to improve the strength and provide drainage along the road corridor.

Acknowledgements

My appreciation goes to my Supervisors, Prof. Olusegun Ige and Prof. Olufemi Ogunsanwo for their support and indebt supervision of the research as well as Nigerian Building and Road Research Institute for assisting in the laboratory test.

Conflict of interest

The authors declare no conflict of interest.

Abbreviations

Grain size distribution parameters according to their diameter

G	gravel size (4.0–13.2 mm)
CS	coarse sand size (0.6–2 mm)
MS	medium sand size (0.2–0.425 mm)
FS	fine sand size (0.075–0.2 mm)
S	silt size (0.002 mm)
C	clay size (0.001 mm)

Atterberg limit parameters

LL	liquid limit
PL	plastic limit
PI	plasticity index
wPI	weighted plasticity index [(PI × % passing 0.425 mm)/100]
NMC	natural moisture content
Fsw	free swell

Compaction parameters

MDD	maximum dry density
DD	dry density
BD	bulk density
OMC	optimum moisture content
MC	moisture content
CBRu	unsoaked California Bearing Ratio
CBRs	soaked California Bearing Ratio

Other derived parameters

LLr	liquidity ratio (LL/MC)
PIr	plasticity ratio (LL/PL)
Wr	moisture ratio (MC/OMC)
Dr	density ratio (DD/MDD)

Ac	activity (PI/clay fraction)
SP	swelling potential (PI/PIr)
SD	standard deviation
CV	coefficient of variation [(SD/mean) × 100]
R	correlation coefficient
R^2	coefficient of determination
RMSE	root mean square error

Author details

Attah Fakeye[1*], Olusegun Ige[2] and Olufemi Ogunsanwo[2]

1 Road Research Department, Nigerian Building and Road Research Institute, Otta, Ogun State, Nigeria

2 Department of Geology and Mineral Sciences, University of Ilorin, Ilorin, Nigeria

*Address all correspondence to: attahfakeye@yahoo.com

References

[1] Patel RS, Desal MD. CBR predicted by index properties for alluvial soils of South Gujarat. In: Proceedings of the Indian Geotechnical Conference. Mumbai: Geotrendz. IGS Mumbai Chapter & IIT Bombay; 16-18 December 2010. pp. 79-82

[2] Carter M, Bentley SP. Shrinkage and swelling characteristics. In: Soil Properties and Their Correlations. Chichester, UK: John Wiley and Sons; 2016. DOI: 10.1002/9781119130888.ch8

[3] Owoseni JO, Adeyemi GO, Asiwaju-Bello YA, Anifowose AYB. Engineering geological assessment of some lateritic soils in Ibadan, southwestern Nigeria using bivariate and regression analyses. African Journal of Science and Technology (AJST), Science and Engineering Series. 2012;**12**(1):59-71

[4] Yildirin B, Gunaydin O. Estimation of California bearing ratio by using soft computing systems. Expert Systems with Applications. 2011;**38**(5):6381-6391

[5] Yitagesu FA, Van Der Meer F, Van Der Werff H, Seged H. Evaluation of soil expansion index from routinely determined geotechnical parameters. Soil Science Society of America Journal. 2011;**75**:1640-1651

[6] Oluyide PO, Nwajide CS, Oni AO. The Geology of Ilorin area with explanations on the 1:250,000 series, Sheet 50 (Ilorin). Geological Survey of Nigeria Bulletin. 1998;**42**:1-84

[7] British Standards Institution 1377: Methods of Test for Soils for Civil Engineering Purposes. General Requirements and Sample Preparation. London, United Kingdom: British Standards Institution; 1990. p. 6. ISBN: 0-580-18030

[8] Jollife IT. Principal Component Analysis. 2nd ed. Springer series in Statistics. New York, USA: Springer-Verlag; 2002

[9] Murtagh F, Legendre P. Ward's hierarchical clustering method: Clustering criterion and agglomerative algorithm. Journal of Classification. 2014;**31**(3):274-295

[10] Ige OO, Fakeye AM, Ogunsanwo O. Geotechnical assessment of some along Ilorin-Lokoja highway—Implication on suitability for road construction. Bulletin of the Science Association of Nigeria. 2018;**29**:149-162

[11] Nwaiwu CM, Afolayan JO, Osinubi KJ. Predicting hydraulic conductivity of compacted lateritic soils, a reliability approach. Electronic Journal of Geotechnical Engineering. 2006;**24**(1):45-56

[12] Adams JO, Aderinola OS, Akinwamide JT. Geotechnical study of pavement indices influencing failures along ado-Ajabandele-Ikere road, South Western, Nigeria. International journal of Engineering Innovation and Research. 2015;**4**(4):567-572

[13] Bayamack JF, Onana VL, Ndzie AT, Ngo'oze A, Ohandja HN, Eko RM. Assessment of the determination of California bearing ratio of laterites with contrasted geotechnical properties from simple physical parameters. Transportation Geotechnics. 2019;**19**:84-95

[14] Adams JO, Adetoro AE. Analysis of road pavement failure caused by soil properties along Ado-Ekiti–Akure Road, Nigeria. International Journal of Novel Research in Engineering and Sciences. 2014;**1**(1):1-7

[15] Ndzie AT, Onana VL, Ngo'oze A, Nyassa OH, Ekodeck GE. Influence

of hydromorphic conditions in the variability of geotechnical parameters of gneiss derived lateritic gravels in a savannah tropical humid area (Centre Cameroon), for road construction purposes. Transportation Geotechnics. 2017;**12**:70-84

[16] Adeyemi GO, Olarewaju VO, Akintunde OB, Meshida EA. Mineralogical and geotechnical characteristics of some subgrade soils in a section of the Ibadan/ Ile-Ife expressway, Southwestern Nigeria. Journal of Applied Sciences. 2003;**6**(2):3536-3547

[17] Johnson RA, Wichern DW. Applied Multivariate Statistical Analysis. Eaglewood Cliffs, New Jersey: Prentice-Hall; 1992

[18] Shukla MK, Lal R, Ebinger M. Soil quality indicators for the north Appalachian experimental watersheds in Coshocton, Ohio. Soil Science. 2005;**169**:195-205

Section 4

Swelling Soil

Chapter 5

Swelling Clay Parameters Investigation Using Design of Experiments (A Case Study)

Yacine Berrah, Serhane Brahmi, Nouar Charef and Abderrahman Boumezbeur

Abstract

The present paper aims to investigate geotechnical parametric effects on the expansion behavior of clayey soils in Tebessa province northeast of Algeria using the Design Of Experiments (DOE) methodology. It has been used as powerful tools based on physical and mechanical properties, data results obtained within laboratory soil mechanics testing. This statistical tool methodology presents the factor screening design to determine the effect of different parameters such as dry unit weight, saturation degree, water content, plasticity index, etc., on the swelling pressure parameter which can be used as expansion behavior of clay indicator. All data previously collected in the studied prone area allows the ability of detailed analysis using design of experiment and parametric optimization process with response surface methodology (RSM). Each variable that present effects on swelling pressure is also discussed. Besides, the obtained models and equations related the factors affecting the expansion process have been determined. At the output process; the response desirability of the screening design methodology can be optimized by maximization or minimization of the optimal values affecting the swelling behavior. This process allows us to find the best describing models, whereas output results may be compared to empirical laboratory tests results to assess the RSM models.

Keywords: design of experiments (DOE), response surface methodology (RSM), laboratory tests, geotechnical parameters, swelling soil, parametric optimization process

1. Introduction

The Design of Experiments theory DOE is detailed and covered in many fundamental books [1–3]; its application to machining studies is discussed by various researches [4–7]. However, they have yet to make any inroads in engineering geology except for environmental engineering areas. In fact, geological and geotechnical engineering researchers, especially those who never heard about it and continue to use different inefficient methods and techniques. Moreover, in DOE there are many commercial software packages as, for example, Design-Expert by Stat-Ease, Minitab by Minitab, R Packages powered by R foundation for Statistical Computing, S-Plus by Mathsoft. A great literature and online sources combined are readily available as commercial software packages that apparently make DOE almost effortless.

Though, the simplicity of DOE is really pseudo-simplicity or masked complexity.

That is, in the first stage of DOE requires the formulation of clear objectives study on the swelling pressure of clayey soils as mentioned in this paper. The statistical model selected in DOE requires the quantitative formulation of the objective(s). A response is considered as objective, which is the result of the process under study presented in **Figure 1**. In satisfying these constraints, the software allowed us to establish minimum criteria for the response variables, then view both feasible and unfeasible regions of specific portions of the design space. The process under study may be characterized by several important output parameters but only one of them should be selected as the response. The response must satisfy certain requirements. It should be the effective output in terms of desirable final aim of the study, also easily measurable, preferably quantitatively and a single-valued function of the chosen parameters (dry unit weight $\gamma_d(kN/m^3)$, water content $w(\%)$, Clay fraction Cf (%), plasticity index $I_p(\%)$, Limite of liquidity (%), Saturation degree (%), the preconsolidation pressure $P_c(kPa)$ and the swelling pressure Ps (kPa) as the output parameter).

Basically, swelling soil experiments is one or a series of tests in which purposeful changes are made to dependent or independent input factors or variables of a system, so we may observe and identify the reasons for changes observed in the output response.

Expansive soil has extensively been found in all over the world and cover especially arid and semi-arid regions, literatures and studies investigate deeply the swelling soils behavior and assume that physical properties, geological facies, mechanical and mineralogical characteristics present the main governs parameters dependency [8–13]. Swelling pressure parameter (Ps) or potential was presented in enormous conducted studies as the indicator of phenomenon that can be used in infrastructure sustainable and geotechnical design [14–19]. The output parameter (Ps) is defined in many ways and depend on the testing procedure, to assess the degree of swell, many procedures including laboratory methods determining swell pressure have been developed by geotechnical researchers and engineers [20–22]. Though swelling pressure methods have been developed by

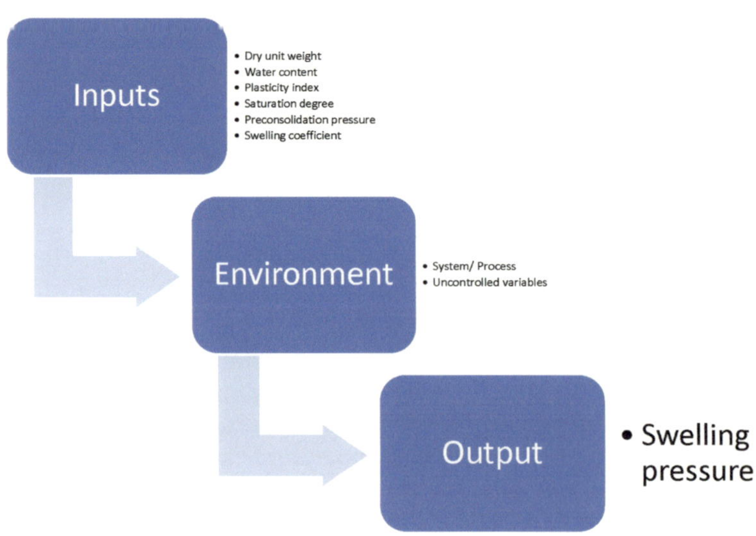

Figure 1.
Visualization of DOE intent.

various researchers, only three methods are standardized and also popularly used as documented in the literature.

The swelling is a complicated phenomenon and the different parameters effects cannot be predictable, used methods for estimating the swelling pressure of clayey soil can be direct or indirect. Direct methods are based on tests, experiences and the basic soil mechanics parameters and provide quick and useful identification, various authors in literature present some empirical relationships with indirect methods [23, 24].

The Tebessa area (Algeria) is the case study of the present work, in point, the weathered geological facies in this arid region has primarily created cover soils in a large basin with very plastic behavior. However, expansive soils exist and well identified litigation and reports high difficulties to infrastructure stability.

In the present research the concept of design of experiments (DOE) has been introduced to study the swelling behavior of the clayey soils with about 121 samples collected and tested in soil mechanics laboratory identification (LTPE).

In various engineering branch, the DOE method is largely used especially in manufacturing and chemical research, it is a powerful approach in experimentations; it seeks to determine the factors affecting a process in relationships with an output of our choice. This research aims to study the swelling pressure as an output parameter affected by several of physical and mechanical parameters as dependent or independent input parameters. Sequential application of DOE plan is used to find the optimal parameter and propose mathematical models to predict the swell pressure generated by clayey soil in Tebessa area and provide recommendations in the quality control measures.

2. Material and methods

The experimentation strategy is an approach to conduct and planning. The best-guess approach, combined, mixture and one-factor-at-a-time approach and factorial experimentation are the main approach used. One-factor-at-a-time for each factor consists of baseline level selected as reference, then varying successively factor in its range remaining and fixing the other factors in the goal to analyze the representative or abstruse factors joint effect on the response.

In this strategy, experiments are conducted by simultaneously varying six factors over two levels (namely low level and high level). The two levels are so chosen that they cover the practical range of the parameters under consideration **Table 1**. This case study presents an example of using the response surface for the modeling of the swelling pressure $\boldsymbol{P_s}\left(\mathbf{kPa}\right)$ and the analysis of results with

Factor	Name	Unit	Level	Low Level	High Level	Std. Dev.	Coding
A	γd	kN/m^3	1.72	1.16	2.06	0.0000	Actual
B	w	%	13.21	11.71	38.24	0.0000	Actual
C	Cf	%	69.31	57.00	98.18	0.0000	Actual
D	WL	%	109.79	36.00	160.00	0.0000	Actual
E	IP	%	54.90	22.00	85.00	0.0000	Actual
F	P	*kPa*	224.52	79.50	270.00	0.0000	Actual

Table 1.
Factors for response surface study.

ANOVA. For the presented implementation of DOE technique, Design-Expert10 software was employed to obtain the appropriate functional equations. The right tools at knowledge of research take in account mathematics and statistics to solve the problem considering each potential of the approximation.

The response surface methodology RSM in DOE techniques is widely used for machining processes. Experiments based on RSM technique relate to the determination of response surface based on the general equation [25]:

$$\mathbf{y} = \mathbf{A}_0 + \mathbf{A}_1\mathbf{x}_1 + \ldots + \mathbf{A}_i\mathbf{x}_i + \mathbf{A}_{12}\mathbf{x}_1\mathbf{x}_2 + \mathbf{A}_{13}\mathbf{x}_1\mathbf{x}_3 + \mathbf{A}_{11}\mathbf{x}_1^2 + \mathbf{A}_{ij}\mathbf{x}_i^2 \quad (1)$$

Where A_0, A_i, A_{ij} are respectively interaction, linear, quadratic and intercept coefficients. x_i input independent variables. Continuous factors affect the quantitative response which is analyzed by response surface methodology (RSM), this later best fitting representative critical factors, commonly chosen in the screening phase of the experimental program. The final obtained results using RSM are polynomial models display the true response surface in the best approximation over a region of factors.

2.1. Definition of the input variables and the output responses

In this study, the effects of input parameters (dry unit weight $\gamma_d\left(\mathbf{kN/m^3}\right)$, water content $\mathbf{w}(\%)$, plasticity index $\mathbf{I_p}(\%)$ Liquidity limits **Wl** (%) Saturation degree $\mathbf{Sr}(\%)$, the preconsolidation pressure $\mathbf{P_c}\,(\mathbf{kPa})$ and the clay fraction **Cf** (%)) on the output response the swelling pressure $\mathbf{P_s}\,(\mathbf{kPa})$. The levels for each factor are tabulated in **Table 1**.

Significant factors are identified using two-level factors as the first technique permit to compare the obtained results in full factorial design, where lower numbers of runs are required in this identification.

Results of the full factorial design are then compared to the results of two-level fractional factorial design, in which much lower number of runs are required to identify the significant factors. Explicitly, the half-fraction design can be also compared to the two-level factorial design.

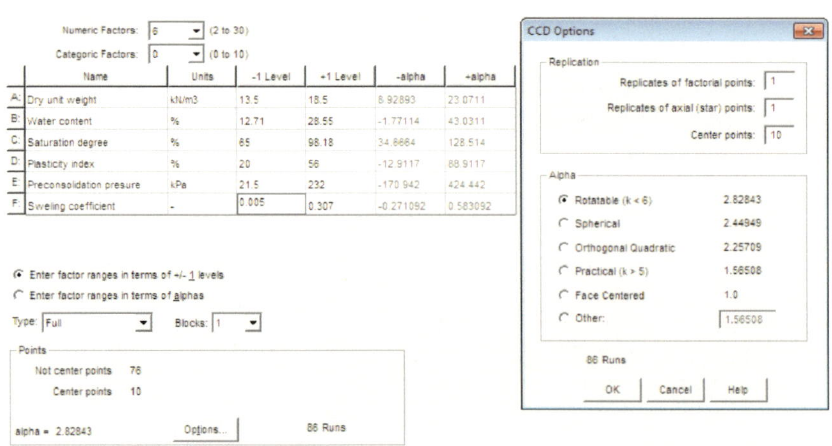

Figure 2.
Definition of different parameters as numeric factors in design-expert, and the dialog box for definition replication points and "alpha" parameter.

2.2. DOE and response data implementation

The two sides unit's length of the cube rang between −1 to +1. "Alpha" is the distance out of cube area measured in levels of coded factors, statistically it is always been a point of discussion view. A variety of Alpha options is presented in the Design-Expert software (**Figure 2**).

3. Statistical results analysis and the model properties

Regression model and test for coefficients significance on individual model achieved using ANOVA method. **Table 2** show summary statistics of the model, values of "Adjusted and Predicted R^2" are higher for quadratic model which is suggested for the present analysis; experimental data analysis was performed to identify statistical significance of the aim's parameters. The dry unit weight, degree of saturation, water content, plasticity index, preconsolidation pressure and the swelling index on the measured response swelling pressure Ps. The model was developed for 95% confidence level with $R^2 = 0.9155$, and the results are summarized in **Table 2**.

In **Table 2** the value of 0.2 between **Predicted and Adjusted R^2** indicate the reasonable agreement.

Adeq Precision is the SNR, greater than 4 is desirable, so the obtained model can be used to delineate a design space.

Std. Dev.	47.60	**R^2**	**0.9155**
Mean	228.75	**Adjusted R^2**	0.8913
C.V. %	20.81	**Predicted R^2**	0.8391
		Adeq Precision	27.2634

Table 2.
Model summary fit statistics.

Source	Sum of squares	df	Mean square	F-value	p-value	
Mean vs. Total	6.384E+06	1	6.384E+06			
Linear vs. Mean	2.082E+06	6	3.470E+05	90.65	< 0.0001	
2FI vs. Linear	**2.048E+05**	**15**	**13653.28**	**5.80**	**< 0.0001**	**Suggested**
Quadratic vs. 2FI	22389.69	6	3731.62	1.65	0.1429	
Cubic vs. Quadratic	1.448E+05	56	2585.61	1.44	0.1180	
Quartic vs. Cubic	68206.19	32	2131.44			Aliased
Residual	0.0000	6	0.0000			
Total	8.906E+06	122	72998.07			

Table 3.
Sequential model sum of squares [type I].

ANOVA for Quadratic model						
Response 1: R1						
Source	**Sum of squares**	**df**	**Mean square**	**F-value**	**p-value**	
Model	2.309E+06	27	85524.15	37.74	< 0.0001	significant
A-γd	17098.00	1	17098.00	7.55	0.0072	
B-W	5782.26	1	5782.26	2.55	0.1135	
C-Cf	7180.22	1	7180.22	3.17	0.0783	
D-WL	684.43	1	684.43	0.3020	0.5839	
E-IP	11086.62	1	11086.62	4.89	0.0294	
F-P	52922.74	1	52922.74	23.36	< 0.0001	
AB	829.09	1	829.09	0.3659	0.5467	
AC	1752.67	1	1752.67	0.7735	0.3814	
AD	414.75	1	414.75	0.1830	0.6698	
AE	12761.47	1	12761.47	5.63	0.0197	
AF	2534.55	1	2534.55	1.12	0.2929	
BC	221.88	1	221.88	0.0979	0.7550	
BD	1146.40	1	1146.40	0.5059	0.4787	
BE	2240.01	1	2240.01	0.9885	0.3227	
BF	244.95	1	244.95	0.1081	0.7431	
CD	4177.66	1	4177.66	1.84	0.1778	
CE	1458.91	1	1458.91	0.6438	0.4243	
CF	2057.89	1	2057.89	0.9082	0.3430	
DE	4307.25	1	4307.25	1.90	0.1713	
DF	403.89	1	403.89	0.1782	0.6739	
EF	10128.29	1	10128.29	4.47	0.0371	
A^2	1219.99	1	1219.99	0.5384	0.4649	
B^2	8073.28	1	8073.28	3.56	0.0622	
C^2	8255.06	1	8255.06	3.64	0.0594	
D^2	1668.27	1	1668.27	0.7362	0.3931	
E^2	2479.02	1	2479.02	1.09	0.2983	
F^2	578.87	1	578.87	0.2555	0.6144	
Residual	2.130E+05	94	2265.96			
Lack of Fit	2.130E+05	88	2420.46			
Pure Error	0.0000	6	0.0000			
Cor Total	2.522E+06	121				

Factor coding is ***Coded.***
Sum of squares is ***Type III - Partial.***
F-value *of 37.74 indicates a significant model. Only a 0.01% chance that F-value could occur due to noise.*
P-values *< 0.0500 implies significant terms model. A, E, F, AE, EF are the chosen terms. Values >0.1000 implies a not significant model term. The other terms may be used to reduce the improved model require to support hierarchy* ***Table 4.***

Table 4.
ANOVA response surface quadratic model, analysis of variance table [Partial sum of squares - Type III].

Select the highest order polynomial where the additional terms are significant and the model is not aliased.

The F-Value of 90.65–5.80 indicates a significant model with P- value <0.0001 that provide the suggested one 2FI vs. linear with 5.80 F-value, out of the cited condition the models are aliased (**Table 3**). In this case A, B, C, BC are significant model terms where P- Values >0.10 as mentioned in **Table 4**.

Normal plot of residuals, shown in **Figures 3–8**, should be in a straight line, in the residuals the errors distribution is normal regards the strait line form. Whereas the nonlinear patterns such as S-curve form implies a non-normality of the error term and can be corrected by a transformation. Residuals versus predicted response should be randomly scattered without pattern as shown in **Figure 8**. Other analysis can be provided in other cases.

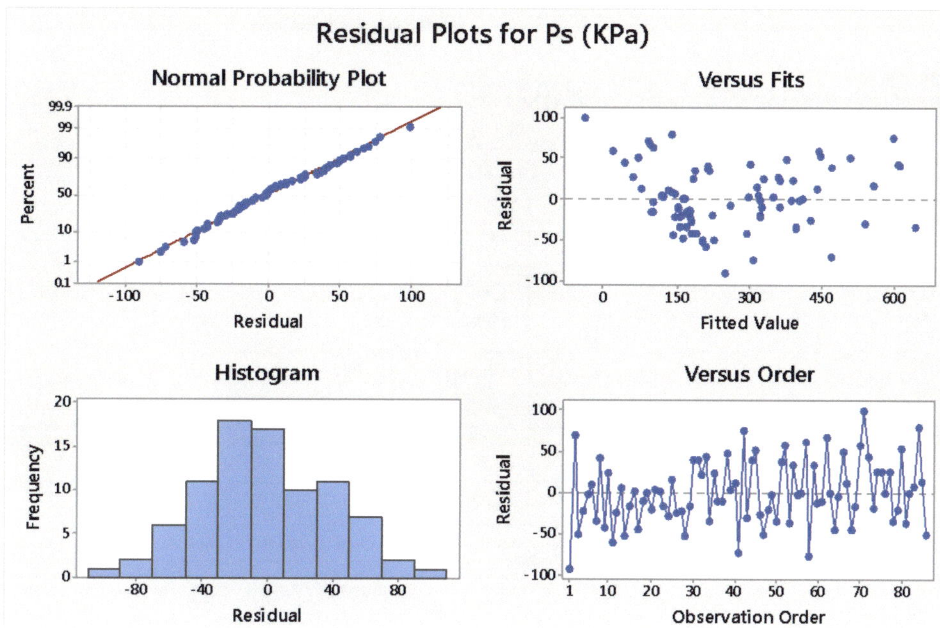

Figure 3.
Residual plots for the swelling pressure of the study soil case.

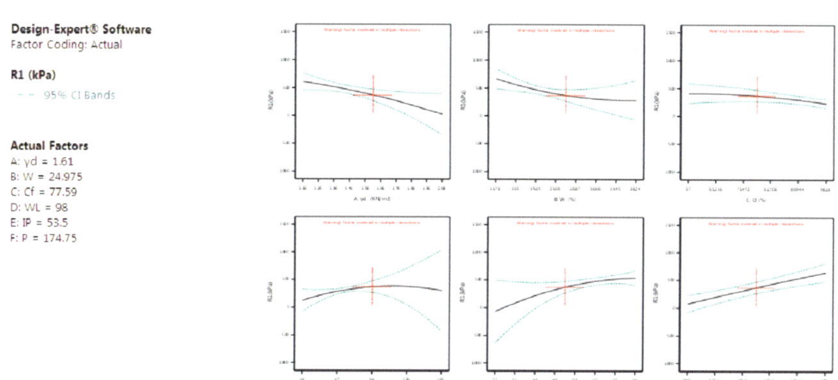

Figure 4.
All factors contribution and effects on the response output for the swelling pressure of the study soil case.

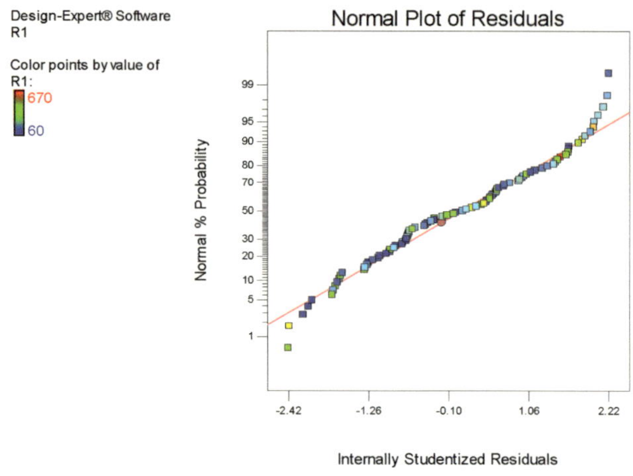

Figure 5.
Normal probability plot of residuals for swelling pressure.

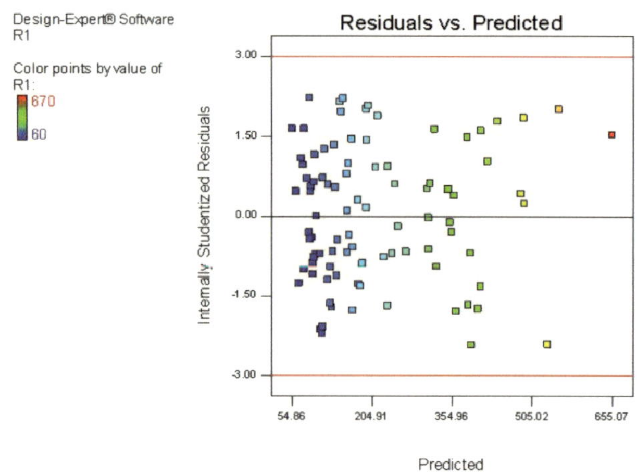

Figure 6.
Residuals versus predicted response for swelling pressure.

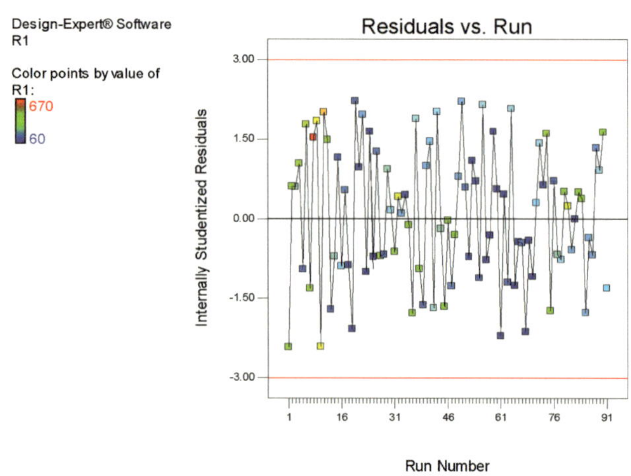

Figure 7.
Residuals versus run for swelling pressure.

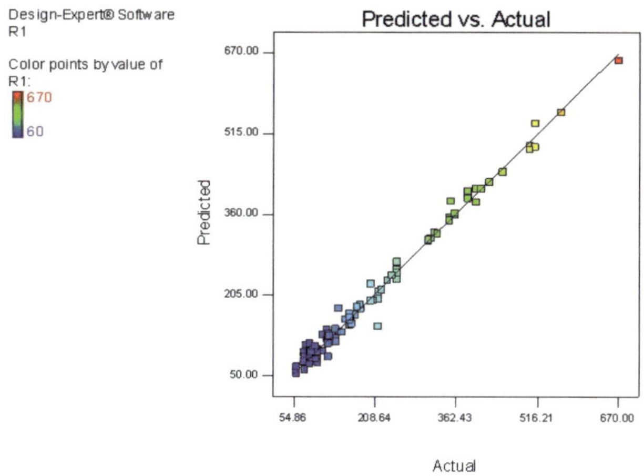

Figure 8.
Predicted response versus actual for swelling pressure.

4. Equations and models graphs

For the analyzed example the final equation in terms of actual factors was determined, which determines the swelling pressure (Ps) from the input factors for the linear model:

$$P_S = 2604 - 10412 * A - 11440 * B + 4034 * C - 2568 * D + 1420 * E + 133 * F \quad (2)$$

And it can be represented by another suggested model of Quadratic form.

The equation can be used to predict the response for each factor levels, that should be specified by their original units. Because the coefficients are scaled to accommodate the factor units and the intercept is not at the center of the design space; this equation is not able to be used in determining relative impact.

Figures 9–11 shows the response surfaces describing the swelling pressure Ps dependence on, the Dry unit weight (kN/m^3) and the water content w (%), plasticity index (%) and Dry unit weight (kN/m^3) and the degree of saturation (%) respectively. Plasticity index (IP) and water content (w) and the preconsolidation pressure, the dry unite weight and the swelling index for this case study.

Figures 12 and **13** represent the factors that affect the (Ps) where the plasticity index is fixed common parameter, saturation degree and the preconsolidation pressure varied respectively.

5. Response surface methodology and optimization process

The response surfaces method is a set of mathematical techniques that use experimental design to determine the range of independent input variables [26]. This method makes it possible, thanks to empirical mathematical models, to determine an approximation relation between the output responses (Y) the swelling pressure $\mathbf{P_s(kPa)}$, and the input variables (dry unit weight $\gamma_d\left(\mathbf{kN/m^3}\right)$, water content $\mathbf{w}(\%)$, plasticity index $\mathbf{I_p}(\%)$ Saturation degree $\mathbf{Sr}(\%)$, the preconsolidation pressure $\mathbf{P_c(kPa)}$ and the Plasticity Index $\left(\mathbf{I_p}\right)$ and limite of plasticity WL) to

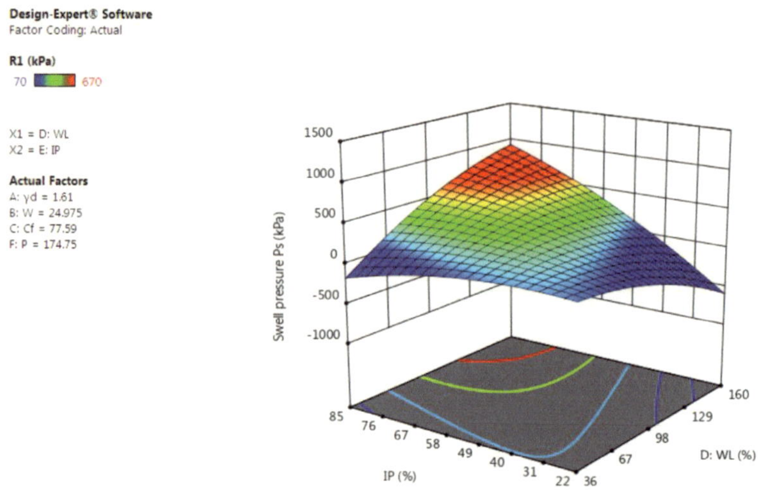

Figure 9.
Response surface 3D representing the swelling pressure dependence on the plasticity index % and the limit of liquidity (%).

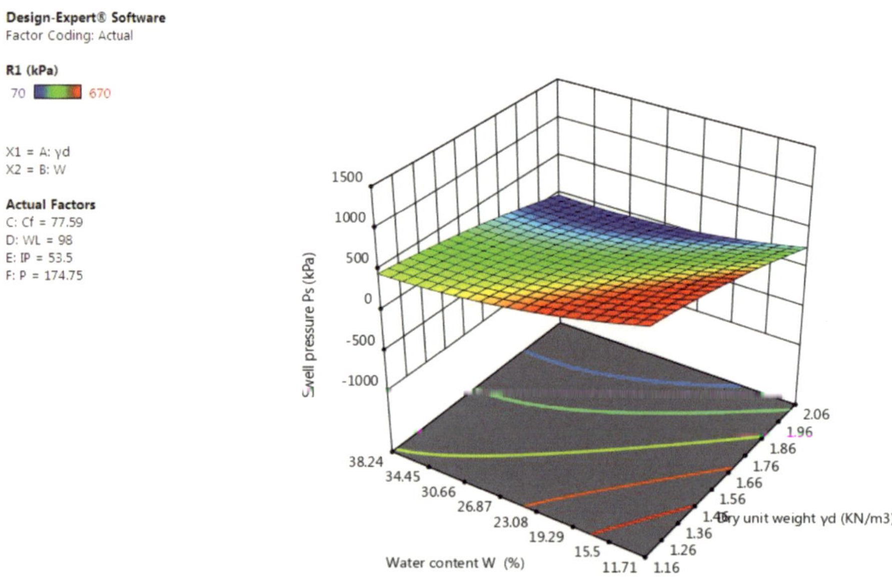

Figure 10.
Response surface 3D representing the swelling pressure dependence on the dry unit weight (kN/m³) and the water content (%).

optimize process parameters to achieve desirable responses. In this method, the answer can be written in the following form:

$$Y = \phi(\gamma d.w.Cf..Ip.Wl.Pc) \quad (3)$$

Where Y is the swelling pressure as the output process and ϕ is the response function, the approximation of Y is proposed using a quadratic mathematical model, which helps to study the interaction effects of process parameters with geotechnical characteristics. In the present work, the second order mathematical model based on RSM is given by the following elements:

$$Y = x_o + \sum_{i=1}^{k} y_i X_i + \sum_{ij}^{k} y_{ij} X_i X_j + \sum_{i=1\,ii}^{k} X_i^{\,2} + \varepsilon_{ij} \tag{4}$$

$$\left(\varepsilon_{ij} = y_{ij} - \overline{y}_{ij}\right) \tag{5}$$

Where x_0 is the free term of the regression equation, the coefficients $Y_1, Y_2, \ldots, Y_k$ and $Y_{11}, Y_{22}, \ldots, Y_{kk}$ are the linear and quadratic terms respectively, while $Y_{12}, Y_{13}, \ldots, Y_{(k-1)}$ are the interactive terms and ε_{ij} presents the fit error for the regression model.

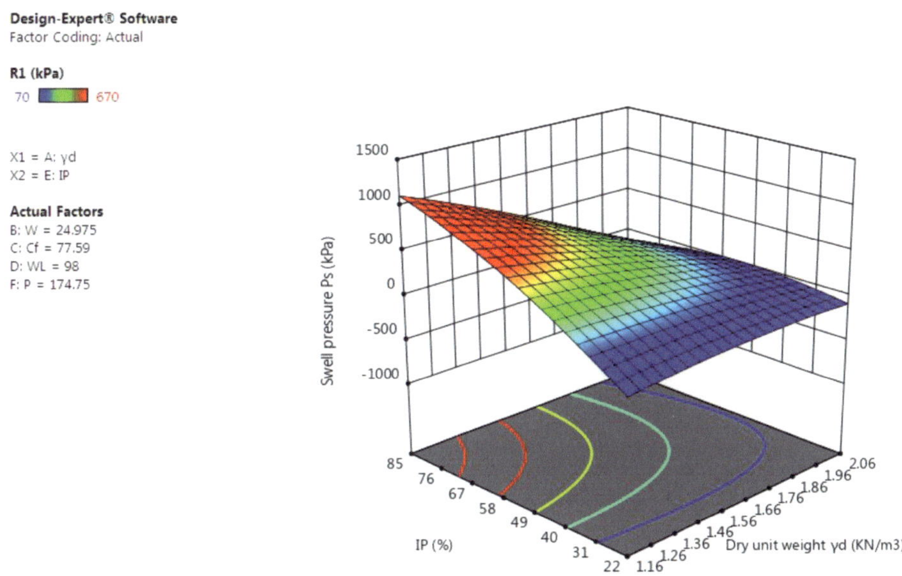

Figure 11.
3D response surface of Ps (kPa) dependence on the $\gamma_d\left(kN/m^3\right)$ **vs IP** *(%).*

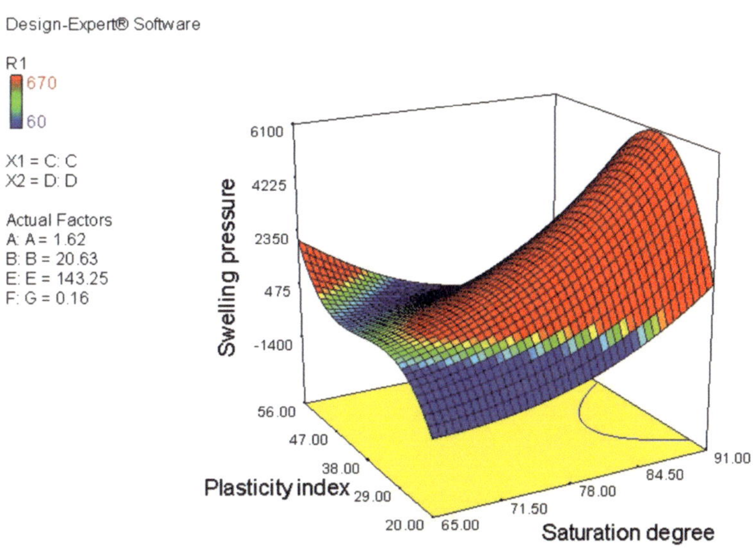

Figure 12.
Response surface 3D representing the swelling pressure dependence on the plasticity index and the saturation degree (%).

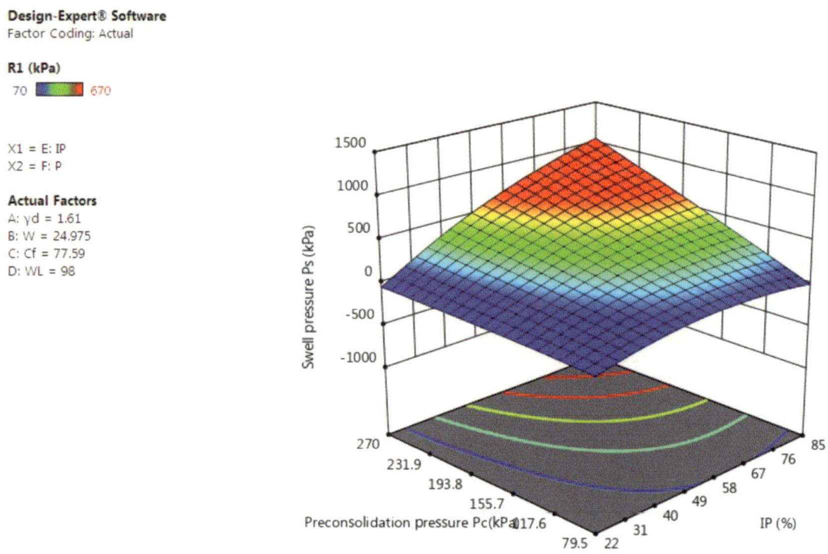

Figure 13.
Response surface 3D representing the swelling pressure dependence on the Preconsolidation pressure and the plasticity index (%).

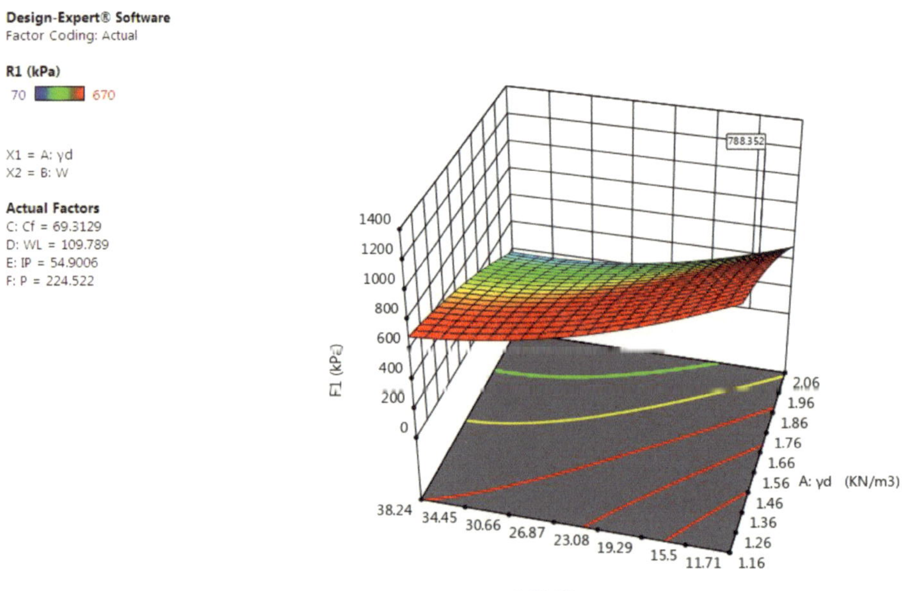

Figure 14.
3D surface of Ps vs. $\gamma_d\left(\mathbf{kN/m^3}\right)$ *and* **w %.**

On the other hand, the coefficient of determination R^2 is defined by the ratio of the dispersion of the results, given by the relationship:

$$R^2 = \frac{\sum(y_i - \bar{y})^2}{\sum(\bar{y}_i - \bar{y})^2} \tag{6}$$

Where y_i: is the calculated response to the ith experience;
$\bar{y}_i$: is the average value of the measured responses.

Analysis of variance (ANOVA) is used to test the validity of the model, as well as to examine the significance and suitability of the model. The model is adequate within a 95% confidence interval. When the values of P are less than 0.05 (or 95% confidence), the models obtained are considered statistically significant. In other words, the closer the R^2 approaches to the value 1, the model is compatible with the real (experimental) values.

3D representation on **Figure 14** clearly optimize the parameters effects on Ps value, based on RSM multifactor data, numerical optimization is possible. Including factors and propagation of error for all variables is available in the settings of Design-Expert software, and limits factor ranges to factorial levels (−1 to +1) in coded values, the area of this experimental design provides the best predictions.

6. Conclusion

Design Of Experiments DOE techniques, using specifically two-levels factorial design method (High and low levels) can efficiently identify the significant factors. Most importantly in this technique is to randomly test at least twice (repeat and replications), in order to reduce the influence of the none assigned variables and the randomness of responses. The present experimentation which was based on six parameters (dry unit weight $\gamma_d\left(kN/m^3\right)$, water content $w(\%)$, Clay fraction Cf (%), plasticity index $I_p(\%)$, Limite of liquidity (%), the preconsolidation pressure $P_c\left(kPa\right)$ and the swelling pressure) on the measured response swelling pressure Ps (kPa) as the output parameter. All parameters varied between 2 levels and revealed that dry unit weight, plasticity index, limit of liquidity, preconsolidation pressure have the main effects on the swelling clayey soil pressure in Tebessa province. The effect of the last factor considered as well as all interaction, to be less or non-significant. The DOE method is most frequently used in simple designs regards to regular fractions, but it does not work as well in more complex settings, such as some nonregular fractions.

Fortunately, the present available general methods work satisfactorily in various situations. It uses a representative polynomial or regression model, by means of one or more methods under the DOE planning analysis and will depend of the user's goal, i.e. if users want a simple analysis, the statistical analysis using the ANOVA approaches can be the ideal method.

In the present research optimization process stage is achieved with response surface method RSM and it revealed that the output parameter (swelling pressure Ps) is strongly affected by plasticity index and liquid limits when desirability is maximized, otherwise the desirability is minimized. In the twice cases the range of all contributed parameters is fixed. Other parameters such as saturation degree show complexed response surface with unclear contribution. Hence the final model of the output response (Ps) do not take in consideration parameters with complex response surface.

Furthermore, the planning of DOE experiments is extremely important in researches because it can reduce cost and time that needs to execute the experimental tests.

Author details

Yacine Berrah[1*], Serhane Brahmi[2], Nouar Charef[3] and Abderrahman Boumezbeur[4]

1 Environment Laboratory, Mining Institute, Larbi Tebessi University, Tebessa, Algeria

2 Water and Environment Laboratory, Mining Institute, Larbi Tebessi University, Tebessa, Algeria

3 Cherif Messadia University, Souk, Ahras

4 Department of Geology, Larbi Tebessi University, Tebessa, Algeria

*Address all correspondence to: yacine.berrah@univ-tebessa.dz

References

[1] Box GEP, Hunter JS, Hunter WG (2005) Statistics for experimenters. Design, innovation, and discovery, 2nd edn. Wiley-Interscience, Hoboken

[2] Montgomery DC (2012) Design and analysis of experiments. Wiley, New York

[3] Maxwell SM, Delaney HD (2003) Designing experiments and analyzing data: a model comparison perspective, 2nd edn. Tylor & Francis, New York

[4] Astakhov VP (2006) Tribology of metal cutting. Elsevier, London

[5] Astakhov VP, Osman MOM, Al-Ata M (1997) Statistical design of experiments in metal cutting—Part 1: methodology. J Test Eval 25(3):322-327

[6] Astakhov VP, Al-Ata M, Osman MOM (1997) Statistical design of experiments in metal cutting. Part 2: application. J Test Eval (JTEVA) 25(3):328-336

[7] Astakhov VP (2012) Design of experiment methods in manufacturing: Basics and practical applications. In: Davim JP (ed) Statistical and computational techniques in manufacturing. Springer, London, pp 1-54

[8] J.-J. Zhang, B.-W. Gong, B. Hu, X.-W. Zhou, and J. Wang, "Study of evolution law of fissures of expansive clay under wetting and drying cycles," Rock and Soil Mechanics, vol. 32, no. 9, pp. 2729-2734, 2011.

[9] ASTM. _2006b_. "Standard test method for expansion index of soils." Annual book of ASTM standards: D4829-03, Vol. 4.08, ASTM, West Conshohocken, Pa., 900-903.

[10] Chen, F.H.: Foundations on Expansive Soils. Elsevier (2012).

[11] K. Terzaghi, "Stability of slopes of natural clay," in Proceedings of the 1st International Conference on Soil Mechanics and Foundation Engineering, vol. 12, no. 3, pp. 161-165, Harvard University, Cambridge, Mass, USA, 1936.

[12] Komornik, A., Wiseman, G., and Ben-Yaacob, Y., "Studies of In-Situ moisture and Swelling Potential Profiles," Proceedings, 2nd International Research and Engineering Conference on Expansive Soils, Texas A&M Univ. Press, College Station, TX, pp. 348-361. (1969).

[13] Fredlund, D. G., Hasan, J. U., and Filson, H. L., "The Prediction of Total Heave," Proceedings of 4th International Conference on Expansive Soils," ASCE and International Society for Soil Mechanics and Foundation Engineering, Denver, pp. 1-17. (1980).

[14] Berrah, Y., Boumezbeur, A., Kherici, N.: Application of dimensional analysis and regression tools to estimate swell pressure of expansive soil in Tebessa (Algeria). Bull. Eng. Geol. Environ. (2016).

[15] Rao, S. M., "Identification and Classification of Expansive Soils," Exapnsive Soils—Recent Advances in Characterization and Treatment, Taylor & Francis, pp. 15-24. (2006).

[16] Jennings, J. E. B. and Knight, K., "The Prediction of Total Heave from the Double Oedometer Test," Transact. S. African Inst. Civil Eng., Vol. 7, pp. 285-291. (1957).

[17] Yilmaz I. Indirect estimation of the swelling percent and a new classification of soils depending on liquid limit and cation exchange capacity. Engineering Geology 85, 295-301. (2006).

[18] Magnan, D., 1993. Caractérisation in situ des sols gonflants : l'essai Expansol.

Thèse de doctorat, Université J. Fourier, Grenoble (France), p 190.

[19] Vijayvergiya V.N., Ghazzaly, G., 1973. Prediction of swelling potential for natural clays. Proceedings, 3rd International Conference on Expansive Soil, Haifa, vol. I, pp. 227-236.

[20] Banerjee, A., Puppala, A.J., Patil, U.D., Hoyos, L.R., Bhaskar, P.: A simplified approach to determine the response of unsaturated soils using multistage triaxial test. IFCEE **2018**, 332-342 (2018)

[21] Pedarla, A.: SWCC and Clay Mineralogy Based Models for Realistic Simulation of Swell Behavior of Expansive Soils (2013)

[22] Pedarla, A., Chittoori, S., Puppala, A.J.: Influence of mineralogy and plasticity index on the stabilization effectiveness of expansive clays. Transp. Res. Rec. **2212**, 91-99 (2011).

[23] KAYABALI, Kamil et DEMIR, Saniye. Measurement of swelling pressure: direct method versus indirect methods. Canadian Geotechnical Journal, 2011, vol. 48, no 3, p. 354-364.

[24] CHEN, Fu Hua. Foundations on expansive soils. Elsevier, 2012.

[25] Myers, R., Khuri, A., & Carter, W. (1989). Response Surface Methodology: 1966-1988. *Technometrics, 31*(2), 137-157. doi:10.2307/1268813

[26] SAID, Khairul Anwar Mohamad et AMIN, Mohamed Afizal Mohamed. Overview on the response surface methodology (RSM) in extraction processes. *Journal of Applied Science & Process Engineering*, 2015, vol. 2, no 1.

Section 5

Water Reservoirs and Dams

Chapter 6

Construction Materials and Dam Foundation While Memve'ele Dam Building in the Craton's Region of South Cameroon

Sylvestre M. Ntomba, Christelle R. Magnekou Takamte, Dieudonné Bisso and Joseph Mvondo Ondoa

Abstract

This chapter mainly focuses on engineering geology for dam construction from the Memve'ele region in Cameroon. Here, it deals with geotechnical and geological proprieties of both construction and dam foundation materials. This study is done at the aim to ensure that these materials need to be improved and how they have been used during dam construction. Field investigations, borehole information, density and seismic velocity measurements have been used, and results indicate that soil deposits have slightly clay content, mechanically well for dam construction and display a weak thickness layer particularly on the dam site. These conditions suggest that soil materials can be used as construction (cushion, transition layers, etc.) and foundation materials after few amendments. Ntem Formations appear weathered and fractured sometimes, though their mechanical behaviors display a good character for civil applications. However, engineering processes have been used to improve it by GIN (Grouting Intensity Number) methods. These formations have been used as construction (rip rap crushing aggregate, etc.) and dam foundation materials. Thus, this chapter contributes to highlight materials and dam foundation conditions which are crucial criteria encountered in the dam with emphasis on both theoretical study and practical application during dam construction.

Keywords: geophysical measurements, geological assessment, geotechnical characters, quaternary deposits, Ntem Formations, Memve'ele dam, Cameroon

1. Introduction

The building for dam required a need for better understanding geological and mechanical properties of investigated sites. This process leads to a comprehensive site characterization and contributes bringing additional work when construction and dam foundation materials need some requirements in order to improve it. These additional works lead maintaining dam stable during his service life. Rock physical and mechanical properties are very important parameters for geological engineering design and construction of dams. Given the number of factors in the designing and construction of dams it can be referred primarily to locally available materials and dam foundation conditions whether these factors fulfilled both geological and geotechnical conditions. Thus evaluation of construction materials and dam foundation must be performed to ensure that these factors are satisfied. Geophysical techniques such as seismic velocity are used to derive quantitative engineering design values in order to assess geotechnical and geological properties during dam site investigations [1–4]. In this chapter, field investigations, borehole logging techniques and seismic velocity measurements are given in order to assess the geological and geotechnical of the Ntem formation and soil deposits that have been used both as construction and foundation materials during dam construction. Both refer to previously published work on Ntem formation characteristics [4] and unpublished data of soil deposit natures in the dam site at Memve'ele.

2. Application of geotechnical and geological results to an engineering assessment of construction materials and dam foundation site

2.1 Geological results

The in situ investigation in the Memve'ele dam site, comprised detail observation, description of geological units observed in the field and in the borehole cores and the laboratory tests include tests for petrographic analysis and on rock samples. The investigation resulted in the geology of dam site display two geological units based on borehole information and surface investigation (**Figures 1** and **2a**) [4]. These two geological units are named quaternary deposits and lower Precambrian Ntem Formations. Sometimes, Ntem Formations emerge and appear highly to moderately weathered down to elevation 386 m below which is slightly weathered diabase. The quaternary deposits are composed of full weathered soil covers with uneven thickness underlain by Ntem Formations in the dam site (**Figure 2a**). Full weathered soil covers with weak thickness occur up to Ntem Formations along the dam axis particularly. The plasticity index (15 < Ip < 40) and the clay content (A-7-5 type) of soil are generally on the high side, but the clay content roughly met the requirement. The subsurface strata are described in the **Figure 3**. The mineral assemblages of Ntem Formations indicate a high grade metamorphicm which corresponds to granulite to amphibolite facies metamorphic conditions. Slightly

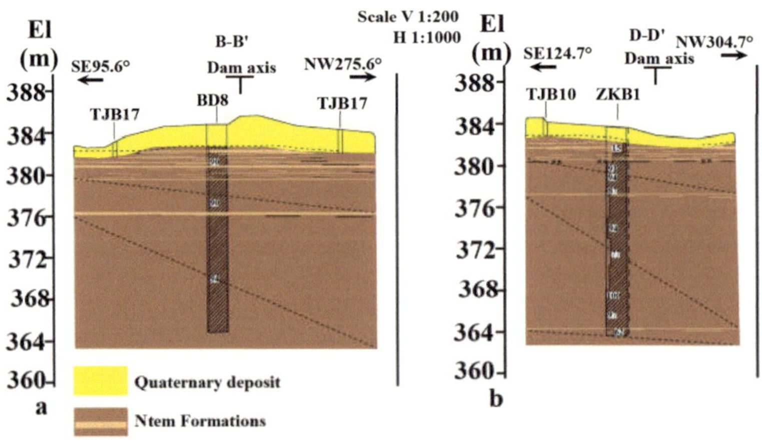

Figure 1.
Geological units in the Memve'ele displaying quaternary deposit underlain by Ntem formation based on borehole information. Note uneven thickness of quaternary deposit between (a) and (b).

Figure 2.
(a) Quaternary soil deposits underlain by Ntem formations that represent two geological units found in the whole dam site and (b) fractured and slightly weathered rock mass.

Shaft Log N°. TJB204

Phase Shaft N°. Depth	Basic Design TJB204 7.10 m	Elevation Coordinate	399.20 m 267183.253 654786.257

Geologic age	Elevation (m)	Depth (m)	Thickness (m)	Graphic Log Scale 1/50	Description of strata
	396.80	2.40	2.40		Silty clay: brownish yellow color, moistish to wet, firm, homogeneous, with plant roots at the top 0.8 m thick.
	393.10	6.10	3.70		Clay: brownish red with brownish yellow color, wet; firm to hard, almost homogeneous 2.4-2.7 m deep with ferruginous nodules and difficult to dig, hard and homogrneous below 3.0 m deep.
	392.10	7.10	1.00		Granitic gneiss: grayish with small brownish red color, completely weathered into debris and sandy material.

Figure 3.
Description of strata.

(**Figure 2b**) and densely jointed zones are found at rocky outcrops, dominantly striking NE30° ~ 40° and E-W with moderate to steep dipping angles (**Figure 4a** and **b**), indicating an outline sub-parallel for all tectonic features (gneissosity, shear zone, fault and jointed rocks). The quantitative description and statistical distributions of discontinuities of Ntem Formations derived from boreholes and those obtained from the geomechanical mapping through scan line at the dam site [5]. The stereographic projection can give an overall view of the number of dominant discontinuity sets of joint and bedding directions within the rock mass. A plot of contoured pole concentration is dominant along the mean direction of N034.44 SE (**Figure 4c**). Quantitative description of discontinuities shows that spacing is extremely close to moderate (0–0.1 m and 30–40 m), their length vary from 1 to 50 m with small

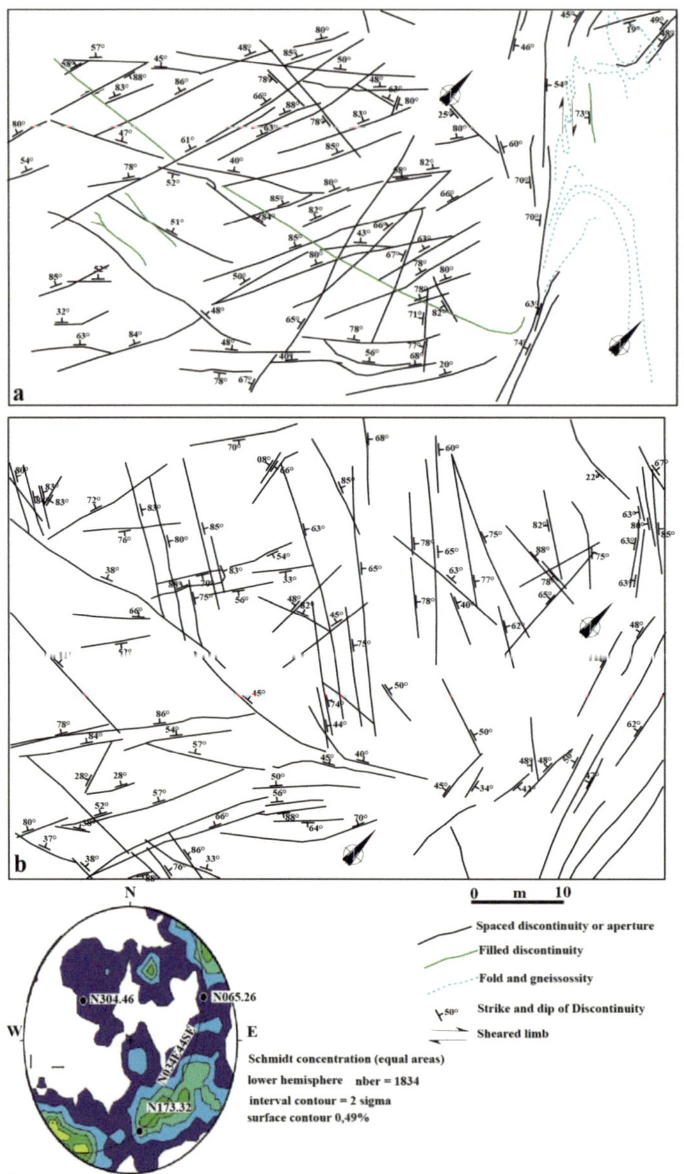

Figure 4.
(a) and (b) Engineering discontinuity map, displaying some discontinuity natures, fold and gneissossity; (c) plot of contoured concentration displaying great circle direction.

DOI: http://dx.doi.org/10.5772/intechopen.93758

Sample N°.	Depth (m)	Physical properties					Atterberg limits			
		Specific gravity	Wet density (g/cm^3)	Dry density (g/cm^3)	Degree of saturated (%)	Porosity	LL (%)	PL (%)	PI	LI
							W_1	W_p	I_p	I_l
Dam site										
TY 07	6.0	2.65	1.60	1.21	72.34	1.200	53.8	36.8	17.0	-0.24
TY 08	2.0	2.64	1.62	1.29	64.86	1.050	51.8	31.0	20.8	-0.25
TY 09	2.0	2.69	1.71	1.41	62.72	0.910	51.4	28.6	22.8	-0.33
TY 10	2.2	2.67	1.64	1.23	75.73	1.170	60.3	33.3	27.0	-0.01
Borrow areas										
TJB101	3.0	2.76	1.65	1.33	61.02	1.070	71.6	38.0	33.6	-0.43
TJB102	3.2	2.71	1.64	1.37	55.14	0.980	58.4	28.3	30.1	-0.28
TJB103	4.0	2.72	1.51	1.26	46.83	1.160	76.2	35.1	41.1	-0.37
TJB204	3.5	2.69	1.75	1.45	65.69	0.860	52.1	27.6	24.5	-0.27
TJB205	3.5	2.69	1.64	1.31	64.03	1.050	83.6	43.5	40.1	-0.46
TJB206	3.8	2.68	1.80	1.46	73.96	0.830	68.0	36.4	31.6	-0.43
TJB402	3.5	2.64	1.56	1.32	47.30	0.990	69.9	32.3	36.6	-0.40
TJB403	3.0	2.60	1.64	1.31	67.15	0.990	74.4	41.2	33.2	-0.47
TJB405	3.2	2.65	1.70	1.37	68.85	0.940	66.4	30.4	36.0	-0.17

Sample N°.	Depth (m)	Consolidation (Sat.)		Quick shear		Coef. of permeab.
		Coef. av$^{1-2}$ MPa^{-1}	Modulus Es$^{1-2}$ MPa	Cohe. C KPa	Fric. Φ°	Vertical K cm/s
Dam site						
TY 07	6.0	0.48	4.8	16.00	33.30	2.96E-05
TY 08	2.0	0.77	2.3	16.00	29.50	2.95E-05

TY 09	2.0	0.95	1.9	14.20	34.70	1.96E−05
TY 10	2.2	0.46	4.4	25.10	31.00	8.62E−04
			Borrow areas			
TJB101	3.0	0.50	4.0	55.30	28.80	4.76E−04
TJB102	3.2	0.66	3.0	23.10	35.10	2.61E−05
TJB103	4.0	0.20	9.9	2.20	41.90	
TJB204	3.5	0.28	5.9	17.00	50.10	9.35E−05
TJB205	3.5	1.00	2.2	14.00	38.00	1.55E−05
TJB206	3.8	0.35	4.8	65.00	38.20	3.73E−05
TJB402	3.5	0.52	3.6	11.10	37.10	4.77E−05
TJB403	3.0	0.26	6.9	40.10	33.10	3.14E−05
TJB405	3.2	0.41	4.8	48.80	21.60	3.32E−05

Table 1.
Geotechnical properties of soil deposits in the Memve'ele dam site.

discontinuity dominants [5]. The dam site area is considered to be passive tectonically and geological disasters are mainly composed by weathering, other phenomena such as landslide, collapsing and debris flow are rare found.

2.2 Geotechnical results investigated on quaternary deposits

A geotechnical investigation into quaternary deposits that have represented construction materials and dam foundation of Memve'ele dam is described in this topic. Many boreholes were drilled by drilling rig to assess the geotechnical of the quaternary deposits from borrow areas and in the dam foundation site. The investigation of three borrows areas and dam sites comprised mainly laboratory tests of soil deposits. The results of mechanical tests are presented in **Table 1**. The specific gravity values vary from 2.60 to 2.76; wet and dry densities are ranged from 1.51 to 1.61 g/cm^3 and 1.23 to 1.55 g/cm^3 respectively; porosity values vary between 0.700 and 1.190%; liquid limit values vary between 58.4 and 76.2 wt. %; plasticity limit values are ranged between 28.3 and 40.1 wt. %. The deformation modulus values vary between 1.9 and 9.9 MPa; the coefficient of permeability varies between 8.62E−04 and 3.35E−05 cm/s. As per the exploration well, the reserve estimations is 182.39 × 10^4 m^3. However, slope occurs with erosion marks (**Figure 5**) after raining seasons.

2.3 Geotechnical results investigated on lower Ntem Formations

A geotechnical investigation into mechanical behavior of Ntem Formations at Memve'ele has been described in detail by Bisso et al. [4]. The rock mass qualities of Ntem formations of the dam site were assessed using sonic well-logging in boreholes and sonic sounding on core samples. Acoustic logging is carried out of from bottom to top of a borehole; the spacing between two receiving detectors and the point distance of detections are 0.2 m respectively. Acoustic test of core consists to penetrate the couple of detector and core by butter. The compressional wave velocity and density results with low values at top and high at the bottom respectively help to interpret the geophysical layers (**Figure 6**). Layer at top is identified as quaternary deposits and layer in the bottom corresponds to Ntem Formations as previously shown in **Figure 2a** and description of strata (**Figure 3**). Thus, the compressional wave velocity and density in the rock mass

Figure 5.
Slope stability affected by raining water.

indicate that there is not only a correlation between velocity and formation layers, but also between rock conditions encountered in the Memve'ele dam site (**Figure 7**), velocities of <5000 m/s indicating highly fractured zone, low density and moderately weathered Ntem Formations with both widely and closely

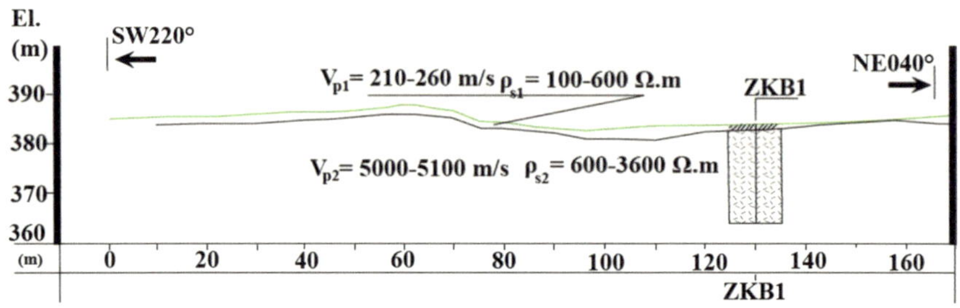

Figure 6.
Seismic velocities, density measurements and geophysical layers in the Memve'ele dam site.

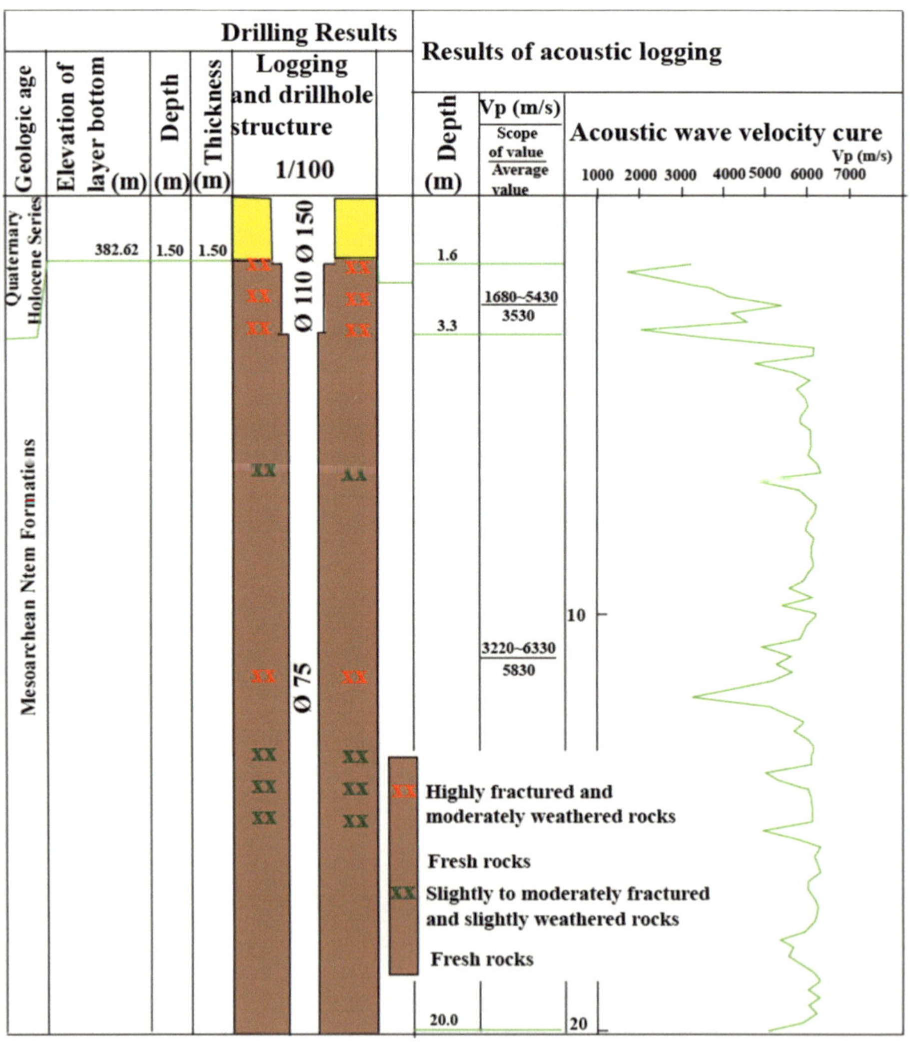

Figure 7.
Drillhole structure and cross plot of seismic velocity measurements.

spaced discontinuities, while velocities of >5000 m/s correlate with fresh to slightly weathered and high density Ntem Formations with very closely spaced discontinuities.

During core drilling, permeability tests were performed and the permeability of the rock mass is expressed in term of Lugeon values. According to Lugeon scale, the Ntem Formations display high, medium and low Lugeon values in the boreholes. However, evidence of two permeability areas is seen and consists to a permeability area with UL > 15, located between 0 and 13 m of depth and an impervious area with UL ≤ 5, beyond of 13 m of depth. The high Lugeon values observed to surface are because of highly jointed rock masses whereas the low Lugeon values correspond to sound rock with moderate fracturation rate. More detailed examination of the competence, softening, density; alkali reactive, hardening of Ntem Formations is displayed by Bisso et al. [4].

3. Interpretations

During Memve'ele dam constructions, geological and geomechanical assessment of quaternary deposits and Ntem Formations have been used to characterize construction and dam foundation materials. It is economically rentable to used construction materials which exist within a relatively short distance of dam site. Interpretation properties on the surface and of measured properties in the borehole logs as mentioned above, display that soil deposits are good to excellent performance in the civil application works [6–9] and thus have been used during Memve'ele dam construction [4]. According to Casagrande scale, soil materials are classified from low plasticity to plasticity and in addition, The coefficient of permeability are higher than those obtained in Kiri dam (from 1.5E−08 to 1.00E−6 m/s [10]) and recommended values of 7.00E−10 to 1.00E−06 m/s. This latter property indicates that soil materials are lesser ability to allow the passage of seeping water if they are use as embankment materials.

The deformation modulus values further indicates the relatively clay contents in the soil deposits. These behaviors indicate that these materials can be used as base layer of dam and also can constituted the transition, cushion and filter materials. The total reserve of soil deposits found near the Memve'ele dam site is 182.39×10^4 m^3 and natural sand is to about 30.000 m^3. These values are more important and contribute to reduce costs if these materials have been brought far of site.

Ntem Formation conditions indicate that they are hard qualities in respect to physical properties and good resistance against scouring. The excavated quantity of rock material from the structure foundation is nearly 300×10^4 m^3 and it has been used for obtaining crushed aggregate and ripping rap materials for dam construction. However, conditions of site with rocks and alluvial deposits indicate very good and relatively poor geological conditions which can constitute dam foundation materials [11]. Concerning dam foundation, evidence of two dam foundation materials is observed and some requirement needs. The thickness of upper residual soil where is weak in the dam axis can be move to obtain Ntem Formations as foundation materials but where, soil deposit is more thickened, it can be used as also foundation materials. This study has shown that Ntem Formations are fractured and weathered. These site conditions have contributed to improve it in excavating the weathered materials and using vegetable cover and Grouting Intensity Number (GIN) [12] for cover slope and introducing the cement grouting through fractures (**Figure 8**). This method leads to reinforce Ntem Formations and slope stability. Overall dam structure can be fitted on the quaternary soil deposits and Ntem Formations.

Figure 8.
(a–c) Display different usages of Ntem formations and quaternary soil deposits; (d and e) amendment of Ntem formations and slope formed by quaternary soil deposits with grouting cement.

4. Conclusion

Geological and geotechnical assessments have been applied to estimate degree of fracture and characterize both construction and dam foundation materials in order to use for engineering geological mapping purposes which can serve as guide while dam construction. The interpretation presented in this chapter shows that quaternary deposits can be used as both embankment materials and dam foundation particularly where his layer is more thickened. The weak thickness of this soil in the island site contributes to emplacement of dam on Ntem formations. Mapping fracture zones and thus weathered grade in the Ntem Formations lead to improve it by engineering processes as a GIN method that leads to improve it. In addition, it has been demonstrated that, cross plots of velocity, density values can be used to predict and identify zones of fracture, zone of weathering, nature and thickness of layers. Previous work [4], cited in this chapter has shown that Precambrian Ntem

Formation based on geological and geotechnical results have been used for engineering design and during Memve'ele dam construction. This approach may have application in other civil applications.

Acknowledgements

The authors express their sincere acknowledges to the Head staff of Memve'ele hydroelectric dam project, Pr. Bisso D., and to the Sinohydro Corporation Limited for access both to the site and the data of the dam. They are grateful to anonymous reviewers for their suggestion to improve the manuscript.

Conflict of interest

The authors have no conflicts of interest.

Author details

Sylvestre M. Ntomba[1]*, Christelle R. Magnekou Takamte[2], Dieudonné Bisso[2] and Joseph Mvondo Ondoa[2]

1 Centre for Geological and Mining Research, Garoua, Cameroon

2 Department of Earth Sciences, University of Yaounde 1, Yaounde, Cameroon

*Address all correspondence to: sylvestre.martial@yahoo.fr; sylmanto.gary@gmail.com

References

[1] Deere DU. Chapter 1: Geological considerations. In: Stagg KG, Ziekiewicz OC, editors. Rock Mechanics in Engineering Practice. New York: J. Wiley and Sons; 1968. pp. 1-20

[2] Hendron AJ. Chapter 2: Mechanical properties of rock. In: Stagg KG, Ziekiewicz OC, editors. Rock Mechanics in Engineering Practice. New York: J. Wiley and Sons; 1968. pp. 21-53

[3] Knill JL. The applications of seismic methods in the prediction of grout take in rock. In: Proceedings of Conference on In Situ Investigations in Soils and Rocks. London: British Geotechnical Society; 1969. pp. 93-100

[4] Bisso D, Ntomba SM, Mengbwa RC, Magnekou Takamte RC, Ondoa JM. Geological and geotechnical characteristics of Ntem formations: Insight of its applications in the Memve'ele dam construction (Southern Cameroon). Geotechnical and Geological Engineering Journal. 2020;**38**:4585-4601. DOI: 10.1007/s10706-020-01312-7

[5] ISRM. Suggested methods for the quantitative descriptions of discontinuies in rock masses. In: Rock Characterization, Testing and Monitoring. Oxford: Pergamon Press; 1981

[6] Paige-Green P, Pinard M, Netterberg F. A review of specifications for lateritic materials for low volume roads. Transportation Geotechnics. 2015;**5**:86-98

[7] Onana VL, Ngo'o Ze A, Medjo ER, Ntouala RFD, Nanga BMT, Ngono OB, et al. Geological identification, geotechnical and mechanical characterization of charnockite-derived lateritic gravels from southern Cameroon for road construction purposes. Transportation Geotechnics. 2017;**10**:35-46. DOI: 10.1016/j. trgeo. 201612.001

[8] Xing H-F, Gong X-N, Zhou X-G, Fu H-F. Construction of concrete-faced rockfill dams with weak rocks. Journal of Geotechnical and Geoenvironmental Engineering. 2016;**132**(6):778-785

[9] Chuka OC, Moruf SB, Ewoma O, Olatunbasun AM. The Kano-Kazaure highway, north Central Nigeria: The significance of the engineering geology in construction. Bulletin of Engineering Geology and the Environment. 2011;**70**(1):33-40

[10] Ahmed Bafeto M, Olugbenga BES, Daffi R, Amina SG. Seepage analysis of kiri dam using finite elements method. The International Journal of Engineering and Science. 2019; **8**(11. Series II):66-86

[11] Fraser AW. Engineering Geology Considerations for Specifying Dam Foundation Objectives. Geology Branch; Division of Safety of Dams California Department of Water Resources; 2001. p. 14. Available from: wfraser@water.ca.gov

[12] Deer D, Lombardi G. Grout slurries - Thick or thin? Issues in dam grouting. In: Proceedings of the Session Sponsored by the Geotechnical Engineering Division of the American Society of Civil Engineers in Conjunction with ASCE, Convention; Denver, Colorado. 1985